The £1000 Racing Car

Building a Competition Car on a Small Budget

Terry Cole

Foulis

Haynes

A **FOULIS** Motoring Book

ISBN 0 85429 482 1

First published 1987

© Terry Cole 1987

Published by:
Haynes Publishing Group
Sparkford, Near Yeovil, Somerset BA22
7JJ

British Library Cataloguing in Publication Data
Cole, Terry
   The £1000 race car! : building a
competition car on a small budget.
   1. Automobiles, Racing—Design and
construction
   I. Title
   629.2'28   TL236
   ISBN 0-85429-482-1

Editor: Mansur Darlington

Page layout: Chris Hull

Printed in England by: J.H. Haynes & Co. Ltd.

Contents

Foreword

by Ian Taylor of Ian Taylor's Race Drivers' School

I got to know Terry Cole when he kept apologizing for how slow he was, on one of my school courses at Thruxton. He may not have been fast, but what he did show was his tremendous enthusiasm for the sport.

Without enthusiasts motor racing would not exist; they are the backbone of the sport, whether they are drivers, mechanics, helpers, marshalls, race officials, medical staff, sponsors or just enthusiast spectators. Terry has written this book for the enthusiast.

Motor racing is an amazingly complex sport, it has many different formulas, classes, categories, clubs, circuits, types of races and so on. Added to this it is constantly changing rules: the rule makers themselves, tyres, engines, chassis and safety features are all subject to continual review and change.

A book like this is essential reading for anyone thinking of entering the sport. All through my racing life I have never been afraid to pick the brains of other people. You have, of course, to learn to sort out the good from the bad advice, but, always be prepared to listen.

As you have already gathered, most people get involved in motor racing for the sheer pleasure and excitement they can get out of this fascinating sport as a hobby. As a driver, no matter how good or bad you may be, or, for that matter, how young or old, if you want to race there is nothing to stop you. Usually, the most difficult restraints to overcome are financial ones, but by reading this book it will tell you many of the ways to go racing on a very limited budget.

For anyone reading this forword who has ambitions to get to the top in motor racing, I will say just a few words. To suggest it is hard is an understatement. You need an abundance of natural talent, you need to have total dedication and, above all you need a fanatical determination to succeed. Apart from all this, you will need to be able to say the right things to the right people at the right time, and luck will have to be on your side – for instance, it helps to have a rich father. But, someone has to be World Champion; it might be you.

As someone who runs a racing school and has raced for many years, the one piece of advice on driving I would give, is this: always remember to look up and well ahead, never look down at the road too much. If you start to have problems with your driving and are perhaps somewhat nervous you always tend to drop your vision; this is a big mistake, avoid it.

Enjoy the book, enjoy your motor racing; I know that you will do both. Thank you Terry for writing this book – I know it will be a great help.

Ian Taylor
Thruxton

Introduction

THE PURPOSE of this book is to introduce you to amateur motor racing and to show you how to build and race a cheap racing car. At the amateur level racing has never been easier, but for many people it appears to be a closed world. This book will open up that world to you.

Motor racing could never be described as a cheap sport, but, by following the instructions within this book, it can come within your reach. Whilst genuine historic racing and sports cars are becoming scarcer, there are a vast number of ordinary vehicles, available at very low prices, with the potential to be rebuilt as racing cars.

You will have the opportunity to race your car at all the famous circuits such as – Brands Hatch, Mallory Park, Silverstone or Donington. There is plenty of space on the grids and the newcomer will be made welcome.

To build your car, you do not need to be a super-mechanic but you must be prepared to do all the hard work yourself. To help you get started I have arranged, with several manufacturers, a discount on some of the specialist items you will need to buy new. I am also offering an information service, so that if you run into problems, or if you have a question which I have not answered in the book, you can write to me. You will find details of these offers at the end of the book in Appendix 3. I have built several racing cars for under £1000, not a few years, but a few months ago, and I actively race the one featured in this book.

Terry Cole

Acknowledgements

Acknowledgements and thanks are due to:

Nick Amey, Andy Anderson, Wendy Markey, the 1984 Classic Saloon Car Champions; Dave Bradley, secretary of the 750 Motor Club, Chris Hart, chairman of the Clubmans Register; Martin Ingal, of the Midget Competition Group; Graham Bayley, Alfa racer; David Walters, Alfa racer; Val Adway, of Performance Services; Chris Winter, friend and fellow racer; J.E. Rossiter, MD of Spax Ltd, R. Hawkins, MD of Jaybrand Racewear; S.C. Hoskins, GM of Fireater Ltd, Stirling Moss, James Weaver, racing driver; Ian Taylor, famous racing driver; Derek Bell, even more famous racing driver. Howard Strawford, MD of Castle Combe race circuit. Gordon Viola, Group 1 Scrutineer, Suzy Livingstone, of the Monoposto Racing Club; Douglas Standley, Capri race series organizer, John Aley, of Aleybars Roll Bars Ltd; Mike Lindsay, of The Alfa Romeo Owners Club; Mike Oxlade, of the British Motor Racing Marshalls Club; Margret and Brian Willis; Jerry Randell, friend and sponsor. Brian Bennet of Victoria Garage, sponsor. *Motoring News; Autosport.* And the following organisations and Magazines HSCC; BARC; BRSCC; BRDS; Mini Seven Club, RAC Motor Sports Association.

1

How to start motor racing

How motor sport is organised

MOTOR RACING is controlled throughout the world by the Fédération Internationale de l' Automobile (FIA), through its competitions committee. The Fédération Internationale du Sport Automobile (FISA). FISA delegates complete responsibility for organizing races to a Motor Sports Authority (MSA) in each member country. In the United Kingdom the designated authority is the Royal Automobile Club Motor Sports Authority (RAC-MSA). It has the power to delegate to various motor racing clubs the authority to organize race meetings.

There are categories within a race meeting which can be classified as follows:

A) International – in which only drivers holding an internationally graded licence may enter.
B) National – in which drivers holding a national or international licence may enter.
C) Restricted – in which drivers holding a restricted, national or international licence may enter.

By far the greatest number of races are organized in the restricted category. All such races are for amateur racing drivers. It is in this category of race we are interested and hundreds of them take place each season.

The racing driver's licence

To drive in any race you will need a racing driver's licence. Such licences are issued by the MSA of the country in which you live. In the UK you should contact: The RAC MSA Ltd, Belgrave Square, London, W1. They will then send you an application form together with a medical report form which will have to be completed by your doctor. Don't panic – you just need all the usual bits in working order plus good vision, with or without glasses, though, naturally, certain medical conditions will debar you from obtaining a licence. The licence has to be renewed each year, for which will be needed a fresh medical examination.

Your first racing licence will be issued for the restricted category only. This means that races are restricted to members of the organizing club or, other clubs that are invited to participate. This helps to ensure that you do not run before you can walk and prevents you entering a race in which drivers of vast experience will be competing for an important championship: you will be in the way and be a danger to everyone.

As a novice driver, you must display on the rear of your car a yellow square with a black X across it. This warns other drivers coming up to overtake you, that you are not experienced and that they should therefore exercise due caution. At each meeting, the organizing officials will sign the back of your licence for each race that you finish. After a certain number of signatures, you can apply to the MSA to have the licence upgraded to a national licence. Further experience and signatures obtained by competing at national category level will qualify you for an international licence.

FIA
(The Federation Internationale de L'Automobile)
Lays down the International Sporting Code

FISA
(Federation Internationale de Sport automobile)
The international competitions committee, deals with all competition matters

Delegates one single club or federation per country recognised by the FIA as the sole international sporting power for control of motor sport in its own country.
Called the ASN.
In Great Britain this is,

RAC
The ASN may delegate control to other bodies and in GB this is the RAC Motor Sports Association Ltd

MSA
The MSA may grant approval for recognised clubs to organise competitions, and to run the 'protected formula' (that is, one specific club is responsible for certain formulas)

The Motor Racing Clubs
organise

International Races	National Races	Restricted Races
These are the ones we are interested in		Competitions organised by invitation (only club members allowed) Competitions confined to vehicles of one make or type.

The organization of Motor Sport

Motor racing clubs

Having been introduced to motor racing and how to obtain a licence, it is now necessary to select a suitable motor racing club to join. It should perhaps be noted that not all 'one make' motor clubs are concerned with racing – for example, the Triumph TR Owners Club, in conjunction with its many other activities, organizes just a few races each year. The clubs which are listed in this book, however, devote virtually all their activities towards motor racing.

You need to select from this list the club which is most suitable for you. This suitability will be based upon:–

A) The type and model of car which you can afford to buy and build.

B) A car which is mechanically simple enough for you to rebuild: for example, if you are an inexperienced mechanic, a Jaguar is much more daunting than an Austin A35. It is interesting to note, however, that Austin A35s regularly win races against Jaguars in circuit races.

All the clubs listed in this chapter offer a very warm welcome to newcomers to the sport; joining the club before you start on the car is a very good idea because, as a member, you will be in contact with drivers who are racing your sort of car. You can go along and have a look at their cars at a circuit before you start on any work. You can see examples of what to do and what not to do: this will save you time and money.

Martin Ingal winning his race at Brands Hatch

Motor racing clubs

750 Motor Club

Contact: Dave Bradley. 16 Woodstock Road, Witney, Oxon, OX8 6DT.
Vehicles: Sports car racing; Formula Ford (for novices); Reliant 750 Formula cars; Formula V; and a whole host of others besides. The 750 Motor Club is a very friendly 'grass roots' racing club.

The Midget Competition Group

Contact: Martin Ingall. South Cottage, Wickhurst Farm, Leigh, Tonbridge, Kent.
Whilst MG Midgets and Austin-Healey Sprites are not amongst the cheapest cars to buy, they are cheap to run and race. Spares and parts are plentiful and inexpensive. The group runs three championships for: road going cars in standard tune; fully modified cars; and a new class for semi-modified cars. The latter class means that is you just want to do simple modifications and perhaps fit wide wheels, you can do so without having to race against fully prepared race only cars.

To enter any of these championships you pay the Midget Competition Group a championship registration fee (about £8), and you must also join either the MGCC or Austin-Healey Club. This is a very fair and reasonable arrangement, since some clubs insist you are a full member of the marque club and separately, of a racing club.

MG Car Club (MGCC)

Contact: Mrs B. Tipton. 67 Wide Bargate, Boston, Lincs.
The above is the address of the national club, there are also seven regional clubs.
Vehicles: any model of MG.

Austin-Healey Club

Contact: Mrs C. Waters. The Laurels, Blind Lane, Tamworth-in-Arden, Solihull, West Midlands, B94 5HT.
Vehicles: any model Austin-Healey. For low budget racing the Austin-Healey Sprite series of cars are of particular interest.

Classic Saloon Car Club

Contact: Mr Peter Defee. 7 Dunstable Road, Caddington, Luton, Beds.
Vehicles: for racing purposes, all pre-1957 saloons, all pre-1965 saloons.
The CSCC is a very important club for the impecunious racer. The CSCC have really taken over the job of providing some forms of low cost racing that many of the big clubs used to provide. Within the age qualifications required of the vehicles there is a vast supply of cheap cars.

British Automobile Racing Club (BARC)

Address: Thruxton Race Circuit, Andover, Hants.
Vehicles: the BARC administer championships for Formula 3, Formula Ford and Formula Ford 2000, and several other professional categories. It also arranges races for Special Saloons, Jaguars (for the Jaguar Owners Club), GTs and MGs, as well as hill climbs and sprints.
BARC also organize marshalling and has a 'marshalls stamp' scheme, whereby you

get cheaper club membership in exchange for your work at marshalling duties. By joining the London and Home Counties regional branch of the BARC, you can compete in an eight-round championship based at Lydden circuit in Kent. Also, very fairly, you do not have to be a full racing member of the BARC to be a racing

member of this group, which saves paying the double fees required by some other clubs. Of particular interest are the championships for Ford Capris and Special Saloons.

You may be thinking that basing a series on one circuit is not much use if you live too far away from the Lydden circuit. This may be true at present, but it is likely that the sport will develop in the future and that amateur race series will soon be available near you as a circuit-based series.

British Racing and Sports Car Club (BRSCC)
Contact: Ms Diane Ellard. BRSCC, Brands Hatch Circuit, Fawkham, Dartford, Kent, DA3 8NH. Tel 0474-874445.
Vehicles: pre-74 Formula Ford, MGs, Duckhams Road Saloon Series, Alfa Romeos, and a host of others.
The BRSCC is another of the 'big three' circuit-based clubs. To enter in the Alfa Romeo-only series of races you must be a member of both the BRSCC and the Alfa Romeo Owners Club.

The Alfa Romeo Owners Club
Contact: The secretary and race co-ordinator, Mike Lindsay. 97 High Street, Linton, Cambridge, CB1 6JT.
Telephone (office hours only) 0223-894300.
Vehicles: any model of Alfa Romeo.
The club has eight regional sections and specialist registers for most of the individual car models. Any Alfa Romeo is eligible for the series of races; there are classes for modified and non-modified cars.

As well as organizing races, the AROC also arranges non-competitive practice days a few times each year at various race circuits; at these you can go to drive round the circuit without all the worry of other cars hurtling past you and this can be very useful experience.

Historic Sports Car Club (HSCC)
Contact: Mr B. Cocks. West Lodge, Norton, Wiltshire.

Telephone (office hours only) 066-63-543.
Vehicles: Historic Sports and single-seater racing cars; Post-Historic Sports cars.

Of particular interest to the racer on a tight budget is the John Lelliot Post Historic Road Sports race series. A vast array of sports cars are eligible for this series and not all of them are expensive cars to buy; for instance the Triumph Spitfire.

British Racing Drivers Club (BRDC)
Contact: H.Q. Silverstone race circuit, Towcester, Northants. Telephone 0327-857271.
Vehicles: a large number of race series are organized by the BRDC along similar lines to the other two big circuit based clubs.

Mini Seven Club
Contact: Peter Tisdale. 33 Stoke Road, Slough, Bucks. Vehicles: all Minis.
The club for all Mini competitions: the one we are interested in being the 850 Mini series. All the notes on racing Mini preparation included in this book, have been supplied by my good friend and fellow racer Chris Winter, who raced Minis with great success before moving on and up, to single-seaters.

Types of races

It is the intention of this book to show you how to prepare a car for circuit racing; all the constructional details, however, apply equally to sprint and hill climb cars.

Sprinting

In a sprint, cars are not competing against each other on the circuit, but against the clock. Each car covers the set course and the car recording the lowest time, in its class, is the winner. Sprint courses consist of a lap of a race circuit or a section of it. There are also specialist sprint courses specifically designed for this sport. For example, that used for the Brighton Speed Trials, which are held once a year on a stretch of the seafront road.

Taking part in a few sprints on race circuits can give you good experience of circuit driving, without having to worry about other cars on the track.

Hill Climbing

A hill climb is a sprint up a hill. It is one of the oldest forms of motor sport and there are venues throughout the country. Racing up a hill requires a different technique and, arguably, a little more skill, than a straight dash round a circuit.

Circuit racing

Circuit racing differs from sprints and hill climbs in as much as drivers are competing against each other in comparable cars. Grid places are determined by the times recorded by each driver in the practice session, which takes place before the actual race.

Amateur races are usually organized into championships, that is, a series of between 10 and 14 races held throughout the season at various circuits. Points are awarded to the winner and other finishers at each race: the champion is the driver who has amassed most points throughout the season. A lot of drivers with limited funds, enter only a few races of a series each year. There are also circuit based championships where a series of races is held at one circuit. If you live within easy reach of such a circuit this can be the cheapest form of competition.

Compulsory protective clothing

As well as holding a racing driver's licence the RAC General Racing Regulations require all racing drivers to wear an RAC approved crash helmet; whilst driving in practice and a race.

There are a large number of approved crash helmets on the market, at prices to suit most pockets. Apart from personal preference and price, the only important point to ensure when purchasing a helmet is that it is on the current RAC approved list (details are in the RAC Blue Book).

An RAC scrutineer's sticker must be affixed to all helmets; these stickers are sometimes already on the helmet at the time of purchase, or will be supplied by a scrutineer the first time the helmet is presented for checking at a race meeting (in this event a receipt showing the date of purchase of the helmet must be shown to the scrutineer).

Fireproof overalls are compulsory for almost all classifications of motor racing. Simple one-piece racing drivers' fireproof overalls are available at reasonable prices; they are also often for sale secondhand (but, see the special offers at the back of this book). More elaborate versions of overalls are available for the fashion conscious. Gloves and socks of fireproof material should also be purchased.

It is most important to wear plain woollen underwear and socks, or Nomex fireproof underwear, underneath the fireproof overalls. Many man-made fibres produce toxins when burnt; when in contact with the skin this can lead to blood

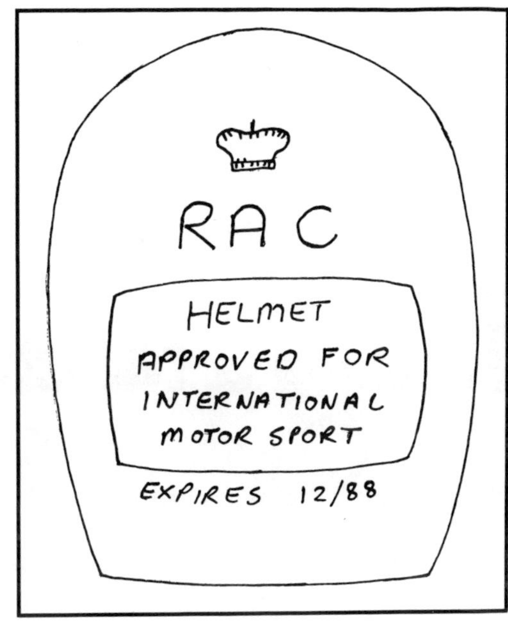

A helmet approval sticker

poisoning. Plain leather boots or shoes, or footwear made especially for motor racing, should also be used.

2

Finding a car and starting work

What sort of car to race

FIRST I HAD better say that you cannot build a Grand Prix Formula 1 car for a thousand pounds – you knew that anyway, didn't you?

You can, however, build a single-seater racing car for £1000; the chapter on single-seaters contains details of these vehicles.

More suitable for a 'first' racing vehicle, however, and one that can easily be built for under £1000, is a sports or saloon racing car – a car that will be safe and exciting, and in which you can enjoy some close competition. It will be a car in which you can learn your racing craft, and one that progresses with your ability – a car that you can realistically afford to run and work on. A car that is the real grass roots of motor racing.

Two points to consider when selecting which type and model of car to build are:-

a) Some cars are only eligible for (or, in a particular state of tune suitable for) one series of races organized by one club.

Example: An unmodified Austin A35 (registered before 1957) is only suitable for the Classic Car Club Classic Saloon series.

b) Some cars are eligible for a number of different series of races.

Example: An Alfa Romeo (registered before 1965) is eligible to enter: Pre-65 Saloon races; Alfa Romeo Championship races; HSCC Historic races.

Chapter and verse on the regulations governing race classes can be found in the RAC 'Blue Book'. This is the racer's Bible and is an absolute must. It contains all the regulations appertaining to racing, what you must have, what you may and may not do to your car, and the lists of what makes and models of car are eligible for what races. The Blue Book is supplied by the RAC with your racing licence, but you can purchase it separately; the price is about £8.00. If you plan to leave applying for your licence until after the car is built; borrow a copy of the Blue Book, but make sure it is up to date.

Where to find a suitable car

Members of the club you select to join will be a good source of information as to the whereabouts of suitable cars, and you will find many know of several vehicles which they have marked down in their diaries. Although drivers keep a note of such vehicles for their own possible future needs, most racing drivers will encourage others and will tell you the location of one of their secret sources and advise you how to obtain the car.

The Magazine *Exchange and Mart* is a much used source, and over a period of a month or so you will probably find therein any car you seek. The magazines *Autosport, Motor Sport, Classic Car,* and *Motoring News* are also good sources for suitable vehicles.

You will often notice in these magazines however, that advertisers are asking unrealistic prices for their cars. If you do see such an advertisement for a vehicle you are seeking you may find that the owner is simply 'living in hope'. It is often worthwhile to contact such an advertiser and make a reasonable offer for the car, giving the owner your telephone number, and asking him to telephone you if the vehicle remains unsold and he has reconsidered the price.

Very occasionally you may find a suitable vehicle in a scrap yard. In such places, what you are looking for is a car with good bodywork and not too much rust. The suspension may be completely wrecked and the engine badly worn, but it will be very cheap. In such condition it will take more work to rebuild, but as you are going to have to take the car almost completely to pieces, it might cost less money at the expense of a little more effort.

Another source for suitable vehicles are

Look for a bargain

dealers who specialise in new and used race cars and parts. These dealers advertise in the magazines *Motoring News* and *Autosport*. Used racing cars are often sold as a 'rolling chassis'; that is, without engine and gearbox. Frequently the owner of such a chassis has built a more modern or faster car, and retained the engine and gearbox to use in the new vehicle. Such a ready-built chassis – with most of the hard work converting it already done – can cost less than building one from a normal road car and save the time involved in working on it.

Graham Bayley in his Alfa GT Junior at Donington

Prices

In order to build a sound, competitive, racing car for under £1000, you should consider a buying ceiling of about £300 for the base car.

In the process of rebuilding the car into a safe racing car, some of the original suspension and other mechanical parts will need to be replaced. Also, nearly all the trim, and seats, will be discarded. For these reasons the cosmetic appearance of the car is not important; accordingly a scruffy, high mileage, but basically sound vehicle, would be a suitable purchase.

A used 'rolling chassis' may initially cost more to buy; however, it may already be equipped with roll bars, racing seat, cut out switch, and fire extinguisher, saving the cost of purchasing these items new. As an example of the budget involved, listed below is my own latest racing car project. The car, a 1964 Alfa Romeo 1300 GT Junior Coupé, was found in *Exchange and Mart.* It boasted a moderate amount of rust, engine and gearbox in pieces, and two large boxes of spares in dubious condition, but was basically sound.

Car bought at,	£250
Set of Spax racing shock absorbers,	£120
Secondhand racing seat,	£30
Set of roll bars and seat belts,	£80
Fire extinguisher, battery cut out switch and sundry small parts,	£60
New parts (bearings, bolts) to rebuild engine,	£100
Welding work, paint, bits and pieces,	£100
Set of tyres,	£100
Total	**£840**

Tools

You do not need a fantastic tool kit to build the car, nor to run it. However, you certainly will need, apart from the usual mechanic's basic hand tools, a comprehensive socket set. Such tools can be purchased at very reasonable prices today, particularly those made abroad. Do not be put off by the sometimes undeserved reputation for poor quality that many of these products have. Good quality tools always cost more to buy; but they do not necessarily do a better job.

Some of the other items you will require can easily be made at home. One such item is a means to remove heavy units, such as the engine and gearbox. A gantry can quite easily be constructed using stout timbers or box-section steel tube. You may need a 'Haltrac' mini hoist or ratcheting puller to lift the engine out; items such as this can be obtained at very reasonable prices at Halfords and similar outlets.

If you need any really specialised tools then you can always consider hiring them. If it is an item such as a compressor to paint a particular part then wait until absolutely everything else for which you can use it is ready; then you can hire it for one weekend and do everything together.

Welding is often a problem for the amateur and unless you have the equipment and skill, or feel there is long-term justification in acquiring both, paying a professional is the cheapest and quickest method of having this work done reliably.

Making a start

Before doing anything else, we ought perhaps, to write out a rough order of work scheme. Working away to complete specific sections of the job in a planned order makes it easier to pull the whole project together at the end. If you start one part, part of another, then put the engine and gearbox in to get it running, you will probably find that they are now in the way of something else you forgot to complete.

Complete each section as nearly as you can, keeping a folder of lists of things you have not done because you are waiting for parts, or welding. Because of working facilities or unsuitable weather, you may well work out a different schedule to the

Car owner Andy Anderson and driver Nick Amey with their championship winning A35

following example; all that matters is that you have some logical scheme to follow. Example:

1.   Strip out cooling system.
2.   Remove engine and gearbox.
3.   Examine bodywork carefully. If it is structurally sound in such parts as suspension points and other important areas, then leave it until last. If there is a wing or two hanging off, this does not matter: you will knock them about anyway working on the car. You can have them welded prior to starting work on making the whole car look neat. If you find some rusted major box section or other structural corrosion or damage, have that attended to right away; you need a sound unit to allow work on the suspension, and later major welding may affect work you have already done.

4.   Once a basically sound car body is achieved you can start work in earnest. Clean all suspension mounting points. Where the underbody, inner wings, chassis sub-frames, are sound, paint them all with a preparation such as 'Finnigans Brown Velvet'. Personally, underseal compounds are not my first choice for a racing car; you may well have to work further on the car during the racing season and a rubber compound gets in your way, clogs nuts and bolts, and has to be stripped for any welding to be done.
5.   Work out what the suspension needs doing to it and get the work completed.
6.   Strip the car interior. Clean off all soundproofing material.
7.   Remove the battery carrier and all extraneous items from the engine bay and boot. Clean and paint these areas.
8.   Pipe all brake and fuel lines through the car.
9.   Lay new battery cables, from the new battery position, inside the cabin or boot.
10.   Sort out and fix roll bars, seat, and seat belts.
11.   Finish all other rolling chassis jobs, such as brakes, fuel tank.
12.   Rebuild engine and gearbox.
13.   Install engine/gearbox unit in car. Check that they work.
14.   Rub down the vehicle paintwork or strip with paint stripper.
15.   Paint the car.
16.   Fix safety stickers, cut out sticker and novice driver plate.
17.   If the car is to be entered in a road-going race class – have the car MOT tested.

3

The body unit

The body unit

MOST OF THE cars suitable for cheap amateur motor racing use will be of unitary construction. In this form of construction the body and all the attachment points for items such as engine and suspension are welded or formed together as a unit. A few cars, for instance the Triumph Spitfire, are constructed with a separate chassis carrying the components. With either type of construction it is important completely to check the unit for corrosion and make repairs where necessary, with particular attention to the suspension and load-bearing areas.

When making repairs to the body/chassis unit, strength and safety are the prime considerations. Where the racing class regulations state that the form of body construction should remain original, safety considerations are allowed to override the regulations.

The body unit should be thoroughly refurbished. All spare trim, and in particular exterior bodywork trim, should be removed. Interior items, such as carpets and soundproofing materials, should be stripped out. Whilst the bodywork and paint finish should be to as high a standard as the budget will allow, the prime consideration is strength, not looks.

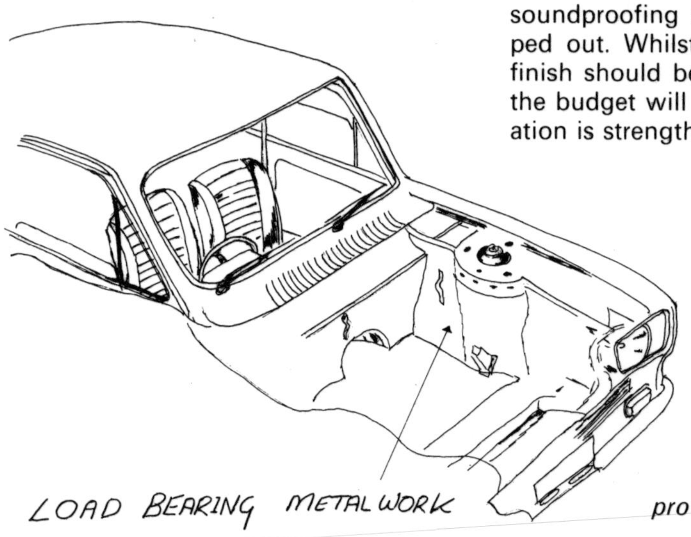

Pressed-steel shape provides the support strength for this Macpherson strut

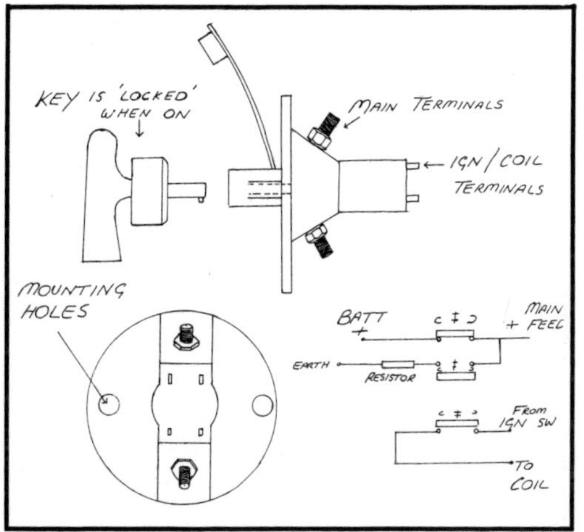

Autolek cut-out switch

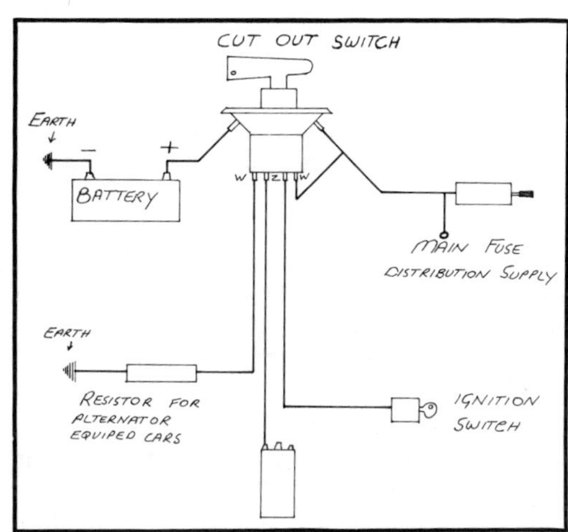

Wiring diagram for cut-out switch

Modifications to the body unit

The following list of modifications for racing use, to the body unit, or components to be fitted to it, are required on virtually all racing cars by the RAC General Racing Regulations.

The safety cut-out switch
The RAC General Racing Regulations state that the cut-out switch must isolate all electrical circuits of the car, and must be fully accessible to anyone from outside the car. The vehicle's normal ignition switch and any ancillary switches, or simplified versions of them, are retained.

On saloon cars the switch should be located outside to the front of the windscreen or just below the rear window. On open cars it should be positioned on the lower main hoop of the roll bar. There must be a sticker on the car bodywork depicting a red spark on a white-edged blue triangle affixed nearby.

The position of the switch is to ensure that in an accident, no matter what the vehicle a track marshall may be dealing with in an emergency, he will know approximately where to find the cut-out switch.

The battery
The battery must be fitted in an acid spill-proof box. The box must be firmly

The battery must be secure in its box

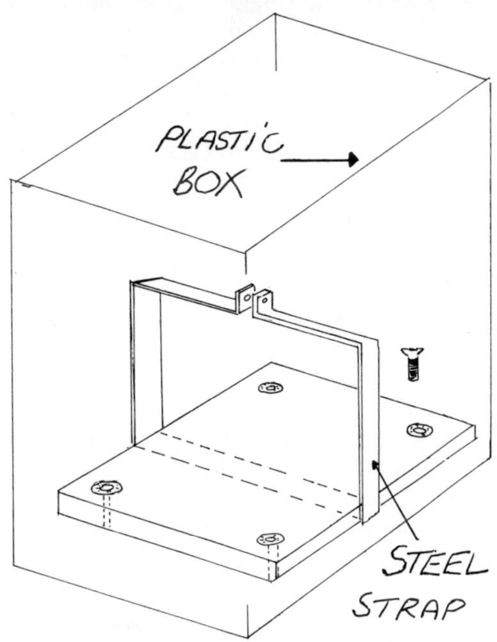

fitted to the car. The battery may be relocated for better weight distribution, for instance in the boot, or to make the engine compartment more accessible.

Good quality heavy-gauge battery cable should be used to rewire a relocated battery: cable of too small a section will cause voltage drop and affect the starter motor performance. The cable should be routed through the inside of the bodywork and securely fixed with cable clips.

Brake pipes

The steel brake pipes distributing the hydraulic brake fluid from the master cylinder to the front and rear brakes is often corroded on older vehicles and should be renewed. Whilst renewing these pipes the rear brake feed line should be rerouted through the inside of the car bodywork. Routing the pipe inside the car – though not mandatory – ensures it cannot be crushed should the vehicle run onto the circuit verge during a race.

Brake pipes can be purchased from many garages made up to length and complete with end-fittings. Halfords sell a cheap tool and the pipe for making brake pipes and end flanges at home. The flexible sections of the brake pipes should be carefully examined and renewed if their condition is in the least suspect. Flexible brake pipes expand slightly with the hydraulic pressure applied; replacing them with 'Aeroquip' hose will give an improved feel to the brakes and a firmer pedal. The hose is available second-hand from race parts dealers.

Fuel line

Although not a mandatory requirement the fuel line should also be rerouted through the inside of the car. The RAC General Racing Regulations state that a fuel line running inside a vehicle must be made of metal, or if non-metal must be enclosed in metal-braided hydraulic pressure hose. Small bore malleable copper pipe sold for minibore central heating systems is suitable. The copper pipe can be covered in braided nylon hose for additional protection.

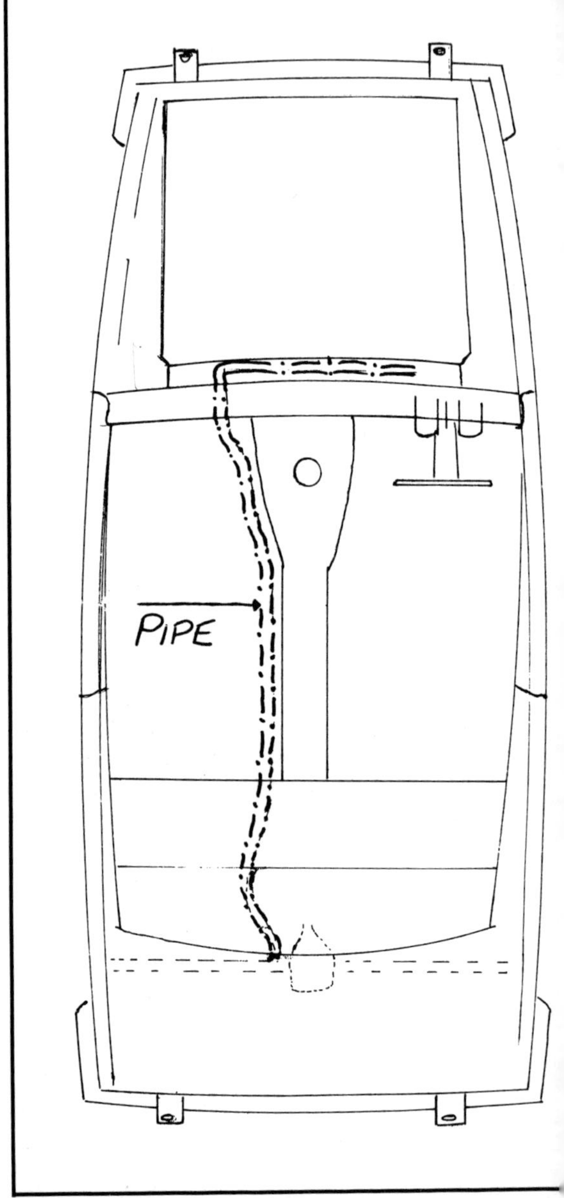

Brake line runs inside car

Fuel pump and fuel filters
The usual mechanical fuel pump fitted to many cars should be replaced by an electric pump. An electric pump should be positioned as near to the fuel tank as is possible since pumps push fuel much better than they pull it.

A good quality large bore fuel filter with replaceable element should be fitted to the fuel line: the Malpassie Filter King is a suitable type. Small in-line plastic fuel filters are too restrictive for racing use.

Fuel tank
For most amateur racing uses the standard fuel tank originally fitted to the car is suitable. The tank should be cleaned and checked for corrosion; then painted with an underseal compound to prevent stones chipping the metal.

The fuel tank breather pipe should have a no-spill valve fitted. Check that the valve operates and renew if faulty.

The fuel tank cap must be a secure fitting: push fit caps are prohibited.

Special racing tanks are manufactured. These tanks, made of aluminium alloys, which bend and crush rather than split if damaged, are fitted with baffles and a foam rubber filling, to combat fuel surge or explosion.

Roll bar shapes

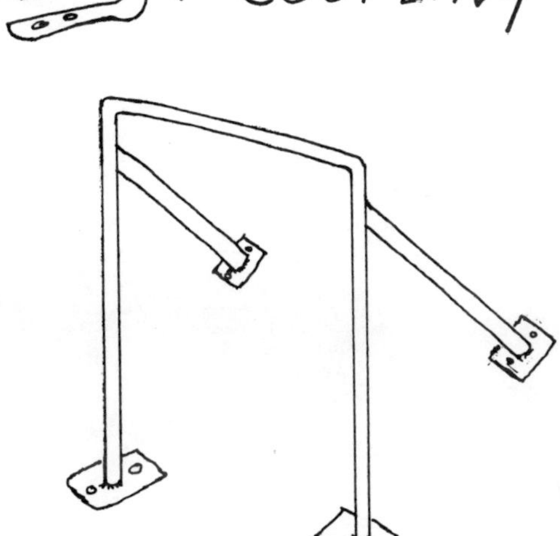

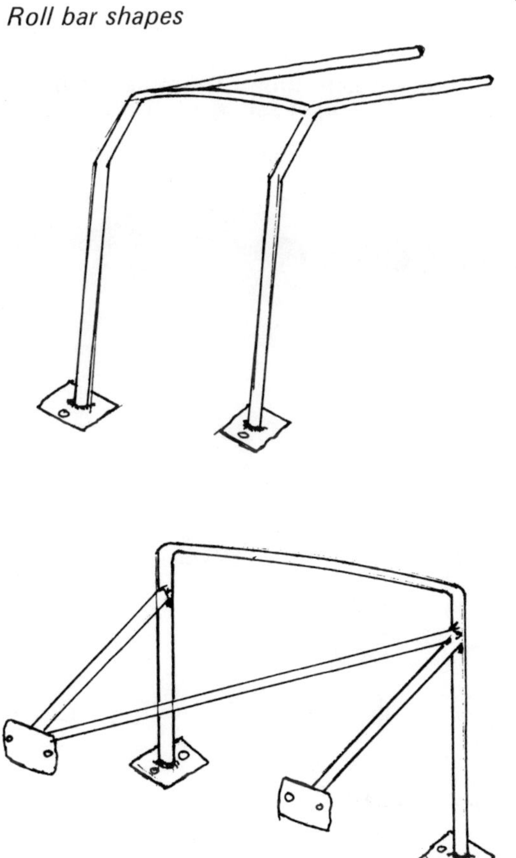

Roll bars
The roll bar is a mandatory requirement for all racing cars. The illustration shows the different roll bar designs available. For light cars (under 1200kg) only a rear roll bar is required.

Roll bars complying with the regulations are available for most cars. Instructions for fitting them are supplied with the bars and simply involves drilling a few holes. Where the bar anchor points are mounted on thin sections of the car bodywork/chassis, such as the floor pan, additional steel plates should be used under the section to ensure the roll bar bolts cannot tear out the metal.

For vehicles where only a rear roll bar is required by the regulations, one can purchase or make a front roll bar for additional protection. The minimum tube size is $1^1/4$ in OD x 16 gauge mild steel; the important dimension is the tube wall thickness: personally I use 10 or 12 gauge for home made bars. The tube must be bent without any crimping deformation on the tube bends. To make the bar an hydraulic tube bender can be obtained from tool hire companies. Do not make too tight a bend in the tube or the wall will flatten. The illustration shows a tube bender with a bar made in four gentle curves, rather than one complete right angle. All roll bars must have a small hole (0.25in) for the scrutineer to check the bar wall thickness.

At any point where the driver's body may

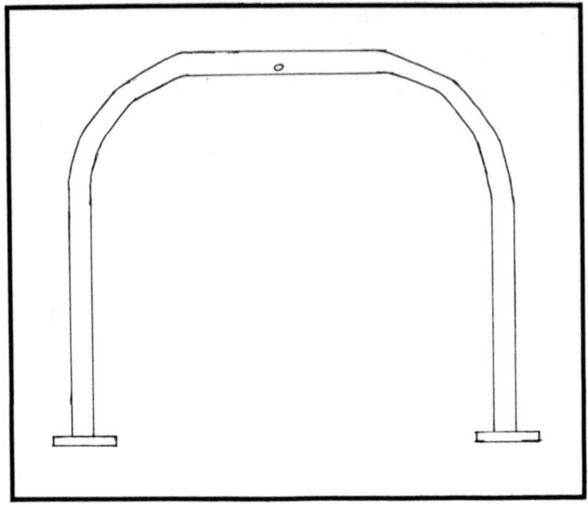

A roll bar with several curves

come into contact with the roll bar the bar should be covered with high-density fireproof foam tube. The foam tube is available from heating and ventilation suppliers.

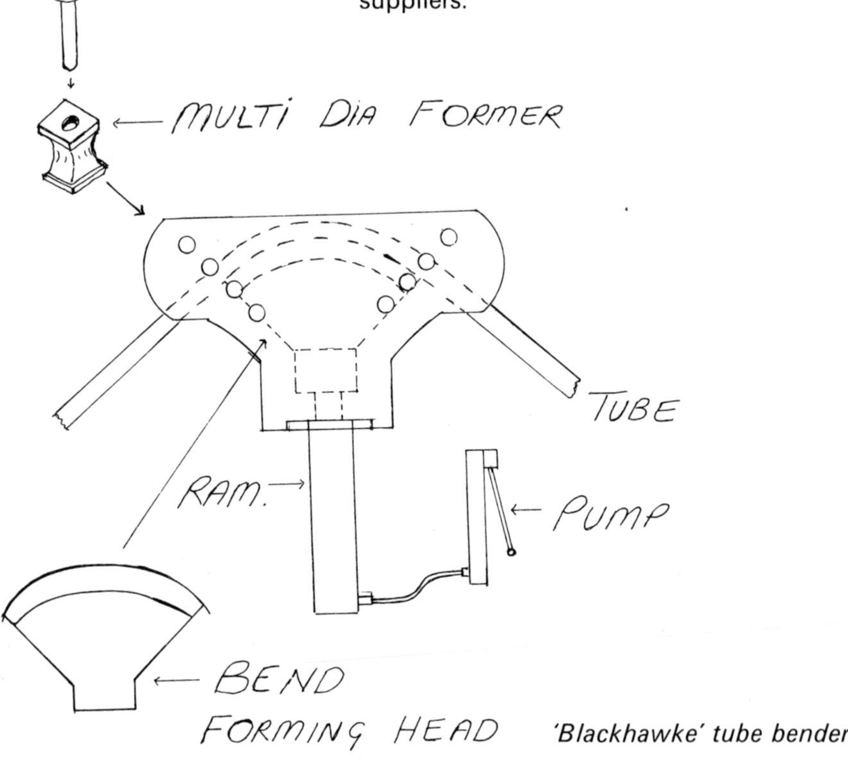

'Blackhawke' tube bender

The driver's seat
A proper bucket seat for racing use should be installed in cars in which the class regulations allow it. Whatever type of seat is used it must be securely fixed to the vehicle floor. Seat mounting bolts which thread into captive nuts in the floor pan should be replaced with longer high-shear bolts which protrude through the floor and are backed up with a substantial steel plate and nut as reinforcement.

Seat belts
In a racing car seat belts are not only for protection but also to locate the driver securely in the driving position: a driver's control of the car can easily be upset by a simple bump into a curb.

The different types of racing seat belt are shown in the illustration. Always buy the best seat belts you can afford. The normal mounting points for the belts on the car should be further reinforced with additional steel backing plates.

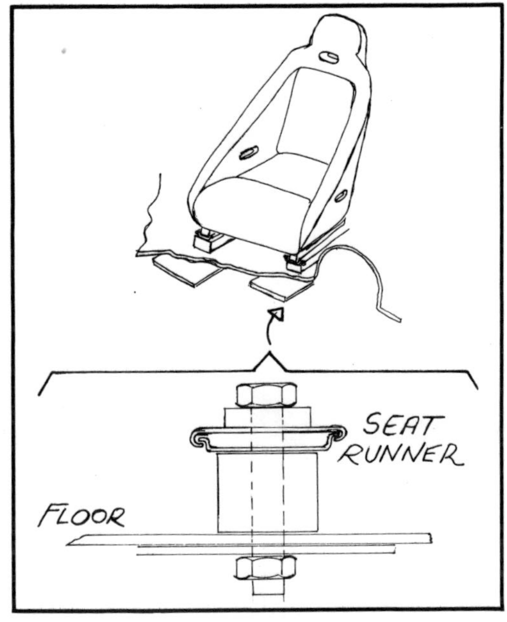

Extra support for a bucket seat fixing

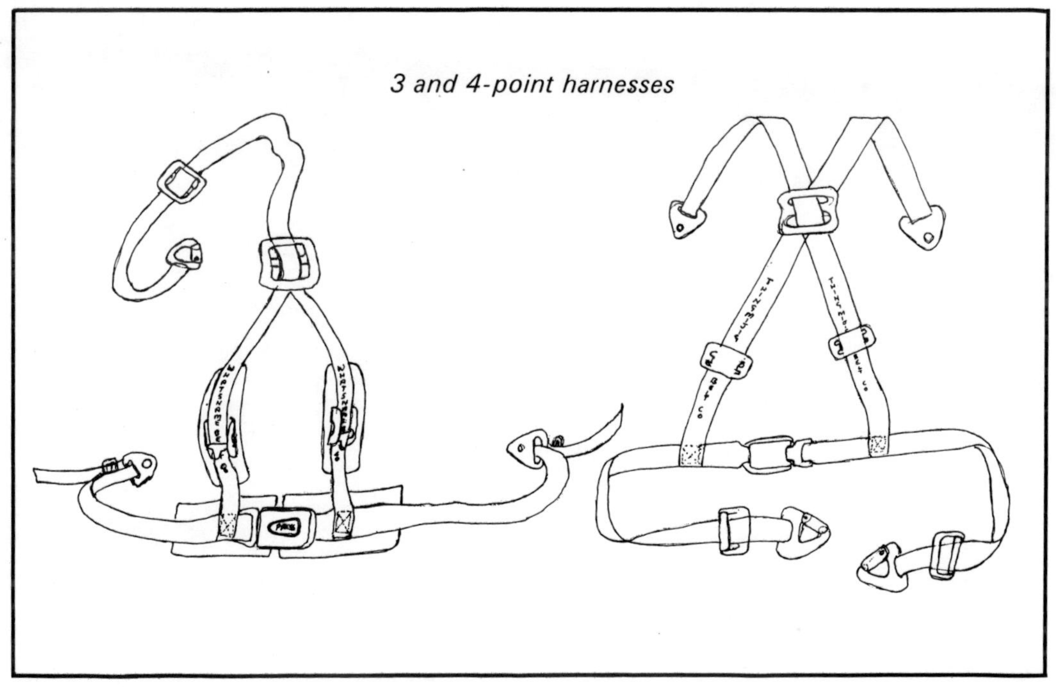

3 and 4-point harnesses

5-point harness

Add plates to seat belt anchors

Crutch strap fixings

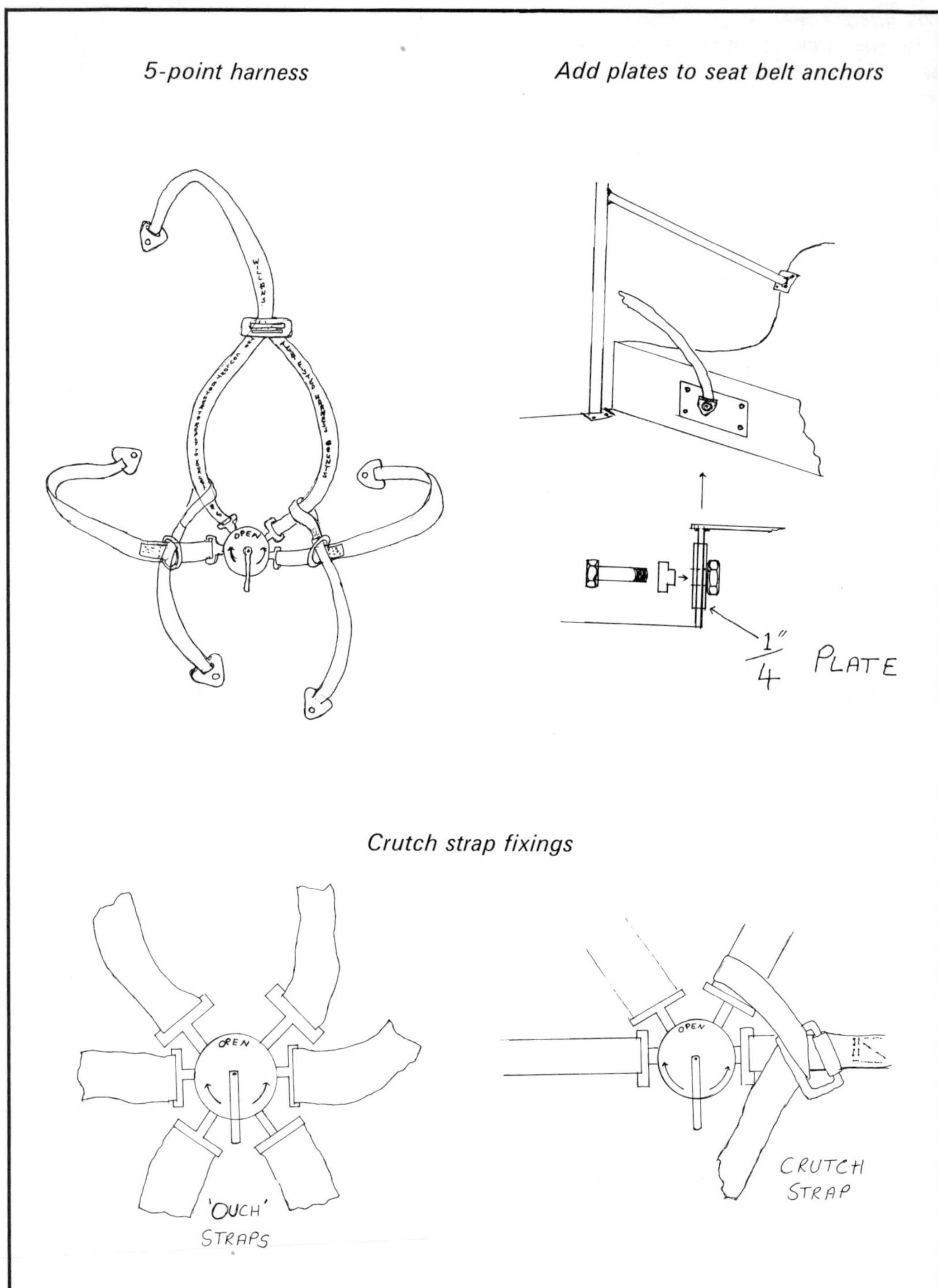

Windscreen and glass
The *only* windscreen materials allowed by the RAC General Racing Regulations are laminated glass or perspex and its derivatives. Tinted glass in any window, which affects through vision, is prohibited.

Windscreen replacement companies can usually supply and fit a laminated screen for most vehicles originally fitted with a toughened glass screen. A scrapyard is also a good source for a used laminated screen.

Plastic screens (which includes Perspex, Lexan and similar materials) can be made at home, but are easily scratched when windscreen wipers are used. To make a plastic screen remove the original wind-screen, tape stout cardboard firmly to it, mark the outline carefully, and cut out the shape on the cardboard. A laminate supplier will cut out the shape for you from the cardboard template.

Firmly tape the plastic screen to the original windscreen and use a hot air paintstripper to heat the plastic until it takes on the shape of the original screen. The same method can be used for the replacement of any of the car windows with a plastic one.

Steering locks
For racing vehicles also used on the road the RAC regulations allow steering locks fitted as original equipment to be left in place. The dashboard should have a small notice affixed stating that a lock is fitted: this is so that track marshalls moving the car in an emergency will know that the ignition key must be turned on.

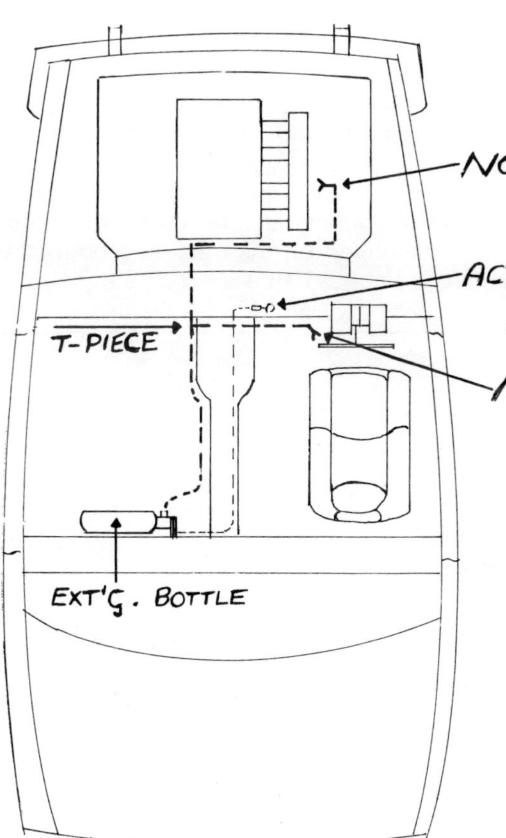

A plumbed-in extinguisher

Fire extinguishers
The RAC General Racing Regulations require a fire extinguisher to be fitted to all racing cars. The minimum size is 1.5 kg (3.3 lb): it must be a BCF or BTM extinguisher.

Extinguishers sold for motor racing use come complete with a secure mounting bracket; this must be firmly bolted to the vehicle.

Extinguishers may be manually operated (which is cheapest) or plumbed in permanently, with an outlet to the driver and engine compartments. Installations using two extinguishers; one for the driver and one for the engine compartment, must be arranged to operate simultaneously.

The catch tank

All racing cars must be fitted with a catch tank for the engine breather pipes and radiator overflow pipes to vent into. The catch tank must be made of clear plastic material so that the contents can be seen. Plastic water bottles used by campers are the most widely used. The tank, which should be emptied after each practice and race, must have a minimum capacity of 2 litres for engines up to 2000cc, or 3 litres for engines over that size. The tank must be firmly strapped in place, and the usual location for it being under the bonnet.

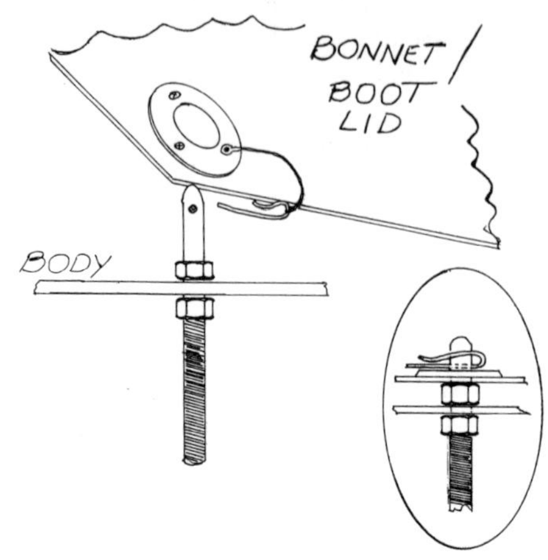

Bonnet pins

Bonnet and boot lid

The usual bonnet and boot lid securing devices should be supplemented by bonnet pins or rubber straps. Whilst motor racing, the vehicle body joints are subjected to higher stresses than are normally encountered on the road, and bonnet and boot catches can easily come undone.

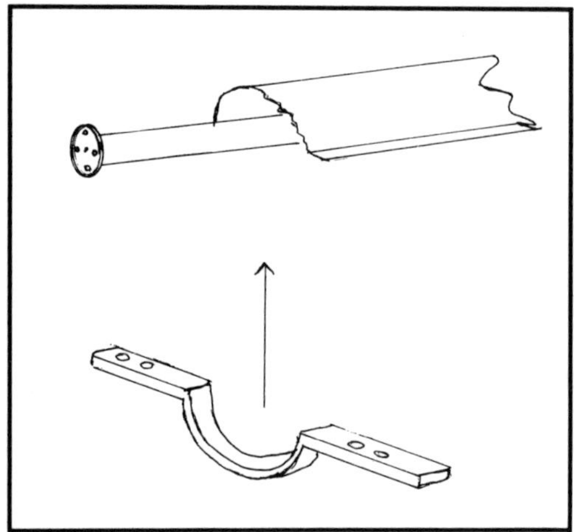

A prop shaft retaining strap

Prop shaft retainer

On all vehicles fitted with a prop shaft a retaining strap must be fitted. The strap is to catch the shaft in the event of a universal joint's breaking. Mild steel strip 1/8in to 3/16in thick is a sufficient size for the strap which should be securely bolted to the floor about midway along the propshaft length.

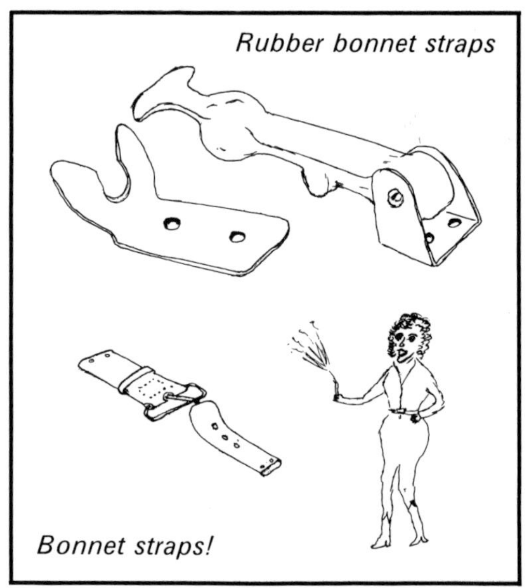

Bonnet straps!

26

Finishing the body unit

All holes in the front bulkhead must be stopped up to ensure a fireproof barrier to the driver's compartment; if the petrol tank is located in the boot the rear bulkhead must also be sealed. Holes can be covered with fibreglass matt and resin, whilst the outlets for cables and pipes should be sealed up with RTV (room temperature vulcanizing) rubber compound.

The standard of body and paint finish is a matter of choice and the budget available for this work: racing cars only need to look 'presentable' and a high standard of finish is not necessary.

Three information stickers are mandatory: a red spark sticker to indicate the cut-out switch position, an on/off direction sticker for the switch, and a sticker indicating the fire extinguisher position. These stickers together with white backgrounds for the race numbers are available from rally shops and similar outlets. Race tape, a wide, coated, linen tape, should be used in crossed strips to tape over headlamp lenses and any breakable light fittings.

Two example cars

Vehicle: Austin Mini 7
Owner: Chris Winter, Southampton
Chris purchased his car as a non-runner with a blown engine: the car had only been used as a normal road car when he bought it. The first job Chris undertook was to remove the engine and gearbox unit. To make it easier to work on the rolling chassis in the confined space of his small garage, Chris then removed the car's doors, bonnet and boot lid, and stripped out all the seats and interior trim; these were stored in a dry shed. To prevent the need for bending, the rolling chassis was then stood on two 18 inch high wooden trestles in his garage.

The finished car was to be fitted with glass-fibre full front section, together with a glass-fibre boot lid. Accordingly, the

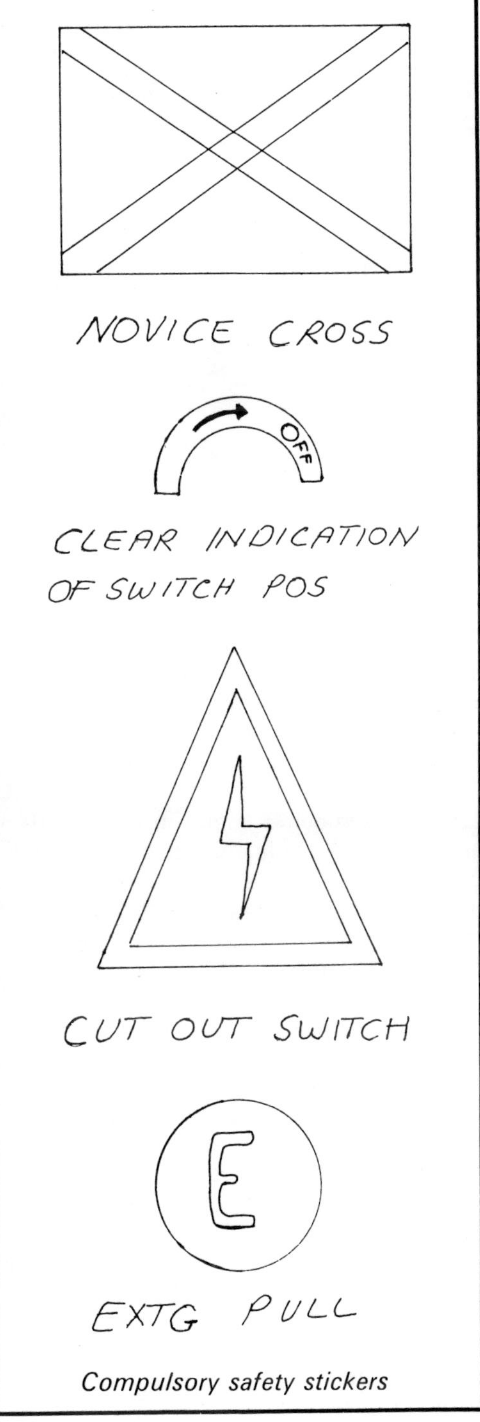

Compulsory safety stickers

Chris Winters' racing Mini

front wings were cut off together with the inner front wings in the area forward of the shock-absorber mountings. To cut the metalwork, Chris used a jigsaw attachment fitted to an electric drill and a strong pair of tin snips.

Whilst removing metal from the front end of the car a hole must also be cut in the bulkhead if a Weber or similar carburettor is to be fitted. Whilst cutting the hole, care must be taken not to cut too near to the front crossmember just below the original speedometer hole. The carburettor hole is finally fitted with a fabricated alloy box extending inside the car; this is required to render the bulkhead fireproof.

Chris recommends tackling the flaring of the rear wheel arches as the next job. The arches must be modified because, with the suspension lowered on the finished car, the rear tyres would foul the wheel arch. The wheel arch is double-skinned at this point, the outer skin being the body shell, and the inner being the wing itself.

There are several ways of carrying out this modification, but Chris uses the following method. A series of radial cuts are made, spaced about 1^1/$_2$in apart, through both the inner and outer wings; as shown in the illustration. Next, each of the 1^1/$_2$in wide tabs formed by cutting the slots, are turned outwards at 90 degrees to their original position. Two layers of heavyweight chopped strand glass-fibre matting are then resin bonded to the outside and inside of the re-formed wheel arch. To finish off, the outer edge of the newly formed arch is trimmed and covered with a proprietary glass-fibre or plastic wheel arch extension.

The final piece of metal removal at the rear of the car was the rear body skirt; this is cut off below the horizontal body flange on which the bumper mounts: the rear skirt adds nothing to the strength of the car.

Chris purchased a professionally-made glass-fibre front end body unit, and the only modification necessary to this for racing use was to enlarge the wheel arches to give clearance for the tyres with a lowered suspension. This involved cutting about 1 inch, tapering down each side, from the top of the wheel arch profile, and making the job neat by fitting wheel arch extensions.

On racing Minis the boot floor can also be removed, which necessitates relocating the battery inside the car. Chris does not recommend this, considering that the weight saved is negligible and preferring to have the strength of the floor retained.

The petrol tank was fitted with an additional strap to stop it moving sideways. The fuel line was then renewed and rerouted through the inside of the car body, with the petrol pump repositioned inside the boot. At the same time the electrical cut-out switch was fitted and wired up. The front and rear bulkheads were then sealed up to provide a firepoof wall to the cabin; all gaps and holes being covered with fibreglass matt and resin.

Chris next tackled the interior of the car, fitting: a roll cage purchased new; a second-hand racing seat; and a homemade alloy dashboard box, in front of the steering wheel, to house the instruments. A 0-10000 rpm rev counter, oil pressure gauge, water temperature gauge, generator and low pressure oil warning lights, were then fitted to the dashboard box, together with light and control switches.

Before refitting the doors, the door trim and pockets were stripped out and the doors strengthened by fitting a diagonal brace made of 1 inch 18 gauge tube. The brace tube can be either welded or bolted into position.

The glass-fibre bonnet and boot lid were fitted; Chris recommends avoiding using the original boot lid hinges. These hinges

The cuts to a Mini rear wing

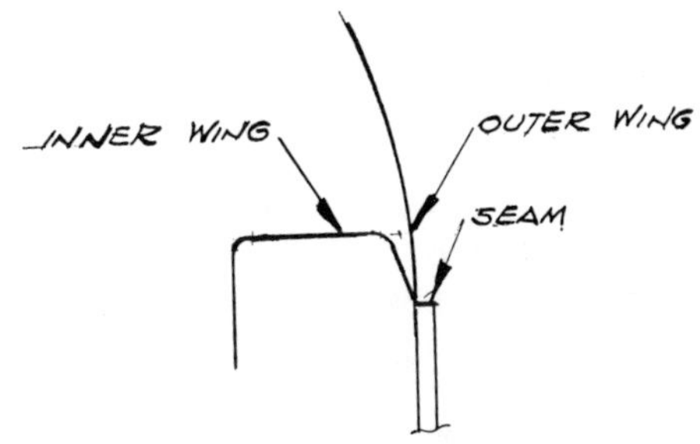

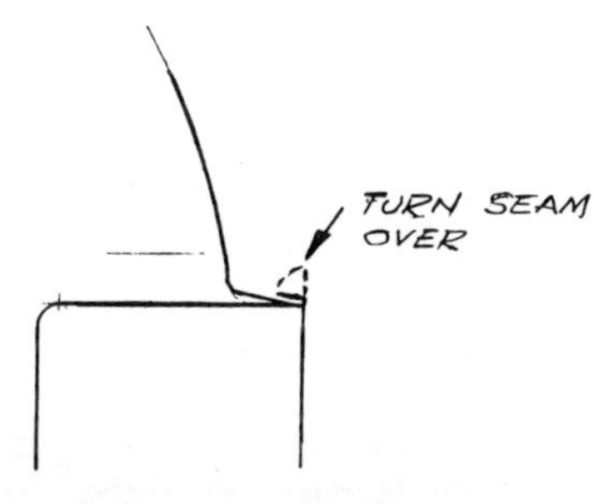

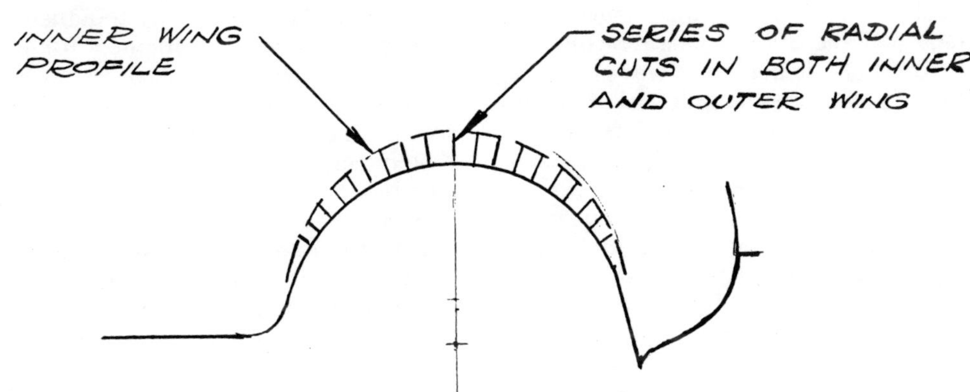

can easily tear out the glass-fibre mould-
ings and spring clips, rubber straps or
bonnet pins are best.

A 13 inch steering wheel was fitted to
the steering column. Chris considers this to
be the ideal size; a smaller one making the
steering too heavy, and a larger one
requiring too much movement of the wheel
on bends. A fire extinguisher was also
installed, on the passenger side floor pan,
within reach of the seated and belted-in
driver.

A laminated windscreen was obtained
and fitted and the side windows were
replaced with perspex ones, cut to shape
by a laminate supplier. The interior work
was completed by painting all exposed
metal matt black. Finally, Chris spray
painted the completed bodyshell in grey
primer, using a cheap airless spray gun. The
final paint finish was left until the suspension had been modified and the engine
unit installed.

*Mini bodies – as can others – can also be
strengthened by seam welding*

Vehicle: Alfa Romeo GT Junior
Owner: the author
This vehicle was purchased in a rather
sorry state of repair and came complete
with several spare panels, doors and
bonnet, from other scrapped vehicles. The
first task was to assess the exact state of
the body unit.

As the engine and gearbox had already
been removed (and came dismantled in
several cases) the car was towed to a
garage and thoroughly steam cleaned. In
the workshop the body unit was placed on
trestles made from lengths of 4 x 4in
timber, supported on axle stands at each
end. The rear axle and all suspension
components were stripped off as was all
the front suspension and steering gear.

Work on the body unit started with
stripping out all the interior: seats, carpets,
soundproofing, except for the headlining,
and the dashboard and wiring loom. Rust
damage was then assessed: it was found
that both sills were badly rusted and that

the box sections inside them were suspect,
that the rear inner wings were rusted away
from the outer wings, and that several quite
large areas of panelling were rusted and
simply tissue paper thin; these were
knocked in with a hammer and the edges
trimmed with snips.

The outer sills were removed using a
hammer and chisel and the inner box
section areas were, indeed, found to be
badly rusted, or missing altogether. New
box sections were fashioned from mild
steel sheet using a block of wood with a
vice and hammer. These were then welded
into position using a Mini Mig welding set.
Commercially made outer sill panels were
purchased and welded over the repaired
box sections.

All the remaining rusted areas were
repaired using mild steel sheet cut to the
approximate shape of the hole and welding
into place using the MIG welder. These
welded areas required considerable grinding down and smoothing out with a small

The author with his Alfa Romeo

angle grinder. Glass-fibre putty was used to fill in and smooth out particularly rough patches left after the welding and grinding.

Inside the car, the area of the rear parcel shelf (which on these Alfas is vinyl-covered hardboard) was covered using a sheet of alloy pop riveted in place: mastic was used on the points of contact under the alloy sheet to ensure a fireproof seal to the boot area.

The brake servo cylinder assembly was moved from under the bonnet (where it takes up a lot of room) and relocated on the floor pan in the passenger side footwell. New brake and fuel lines to the rear were then fitted inside the cabin. The brake line exited to the rear axle through the heel plate of the rear seat position metalwork, and the fuel line entered the boot across the rear seat pan and up and over the inner rear wheel arch.

An electric fuel pump was fitted on the floor of the boot and wired to a switch on the dashboard. Whilst working on these areas a plastic battery box was fitted onto the boot floor and the master cut-out switch installed on the boot lid surround panel, just below the rear window. New battery cables were then installed, with the power feed to the starter motor being run through the cabin interior.

Since the Alfa falls below the general regulations' weight limit requiring a full roll cage, only a rear roll bar unit was installed. This was a secondhand unit originally manufactured to fit an Opel Rekord (or equivalent Vauxhall); with the roll hoop legs shortened by 4in, it was a perfect fit.

A 12inch steering wheel was fitted (using a boss manufactured by Mountney) and a secondhand racing seat installed. To complete the cabin a full harness was fitted with the twin shoulder straps located on a strengthened area of the rear parcel shelf. A fire extinguisher was mounted on the passenger side of the propshaft tunnel alongside the driving seat.

Finally, the entire underside of the car, under the wheel arches, and the interior metalwork, was painted with Finnigans Brown Velvet paint. To protect the newly installed sill box-sections inside the sills, holes were drilled in the sill door plate and the cavities injected with waxoil compound.

4
The suspension

THE TERM suspension is used as a general description not only for the springs of the vehicle but also the axles, arms, steering gear, and devices used to locate the axles and suspension.

The suspension design of road cars is always a compromise between handling and comfort; biased more in one direction than another depending on the make and model.

Virtually all standard passenger road cars are designed to understeer to some degree when cornering; particularly when nearing the limits of suspension travel and tyre adhesion to the road.

Understeer is the term used to describe the condition where the vehicle's front wheels lose adhesion to the road surface and slide before the rear wheels do. Oversteer refers to the rear wheels breaking away first. The aim of modifying the handling of a car is to tune the suspension so that the vehicle understeers less: this will make the vehicle more responsive to a change in direction of travel. Other improvements, such as controlling body-roll, will also enable the car to be driven round corners at greater speed.

Two simple modifications will achieve an improvement in the handling of any car: by making the suspension firmer and fitting racing shock absorbers, which will make the car understeer less with reduced body roll; and by lowering the ride height of the car overall, which will lower the car's centre of gravity and reduce body-roll.

General renovation

First, check the formula regulations for your race class and car to ascertain what, if any, modification to the suspension is allowed.

Racing a car in competition, often driving around corners at maximum speed, places a great strain on suspension components. Track rod ends and ball joints in particular are items that should be regularly examined and renewed in normal use: such items are often overlooked until they fail completely.

Regardless of any suspension modification the suspension should be completely dismantled and rebuilt, any worn parts being replaced.

Study the workshop manual for the model of car and locate all the suspension nuts and bolts that must be undone, then soak them with a freeing oil, such as WD40. These nuts and threads will have been subjected to stress and rust, and it is advisable to replace as many as possible with new ones.

Occasionally bolts or studs may shear off; they can be removed by drilling out and inserting a stud extractor. Component studs can sometimes be replaced with a nut and bolt right through mount and component.

With the suspension dismantled carefully clean around the body/chassis suspension mounting points, and examine the body sheet metalwork in these areas. Important areas are for instance, the body top mounting point of coil springs and the body/chassis rail mountings of leaf springs. Rusted metalwork in these areas should be reinforced with new sheet metal welded into place.

Lowering the ride height

Coil springs. The illustration shows the spring to be a simple coil; however, the spring designer can influence the behaviour of the spring by varying the space between the coils over the length and

Stud extractors

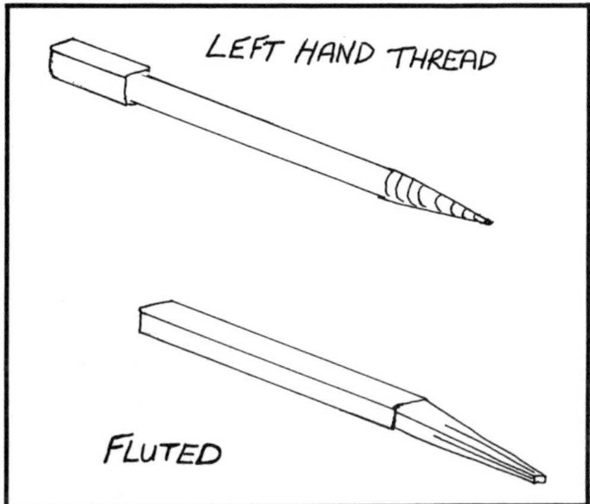

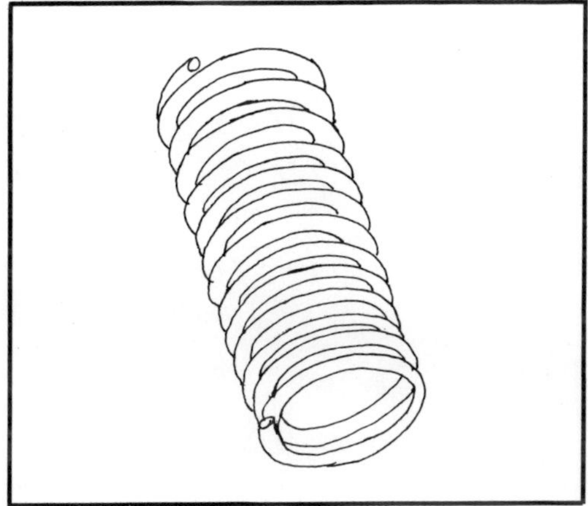

Coil spring

springs of greatly different 'poundage' can be made to the same external dimensions. The coil spring cannot locate the item it provides suspension for and so additional location devices, such as radius rods, are required to hold the axles in place. The coil sprung car can be lowered simply by cutting off one or more coils of the springs. Shortening coil springs increases the 'spring rate', making the suspension firmer, whilst the shorter length lowers the vehicle ride height.

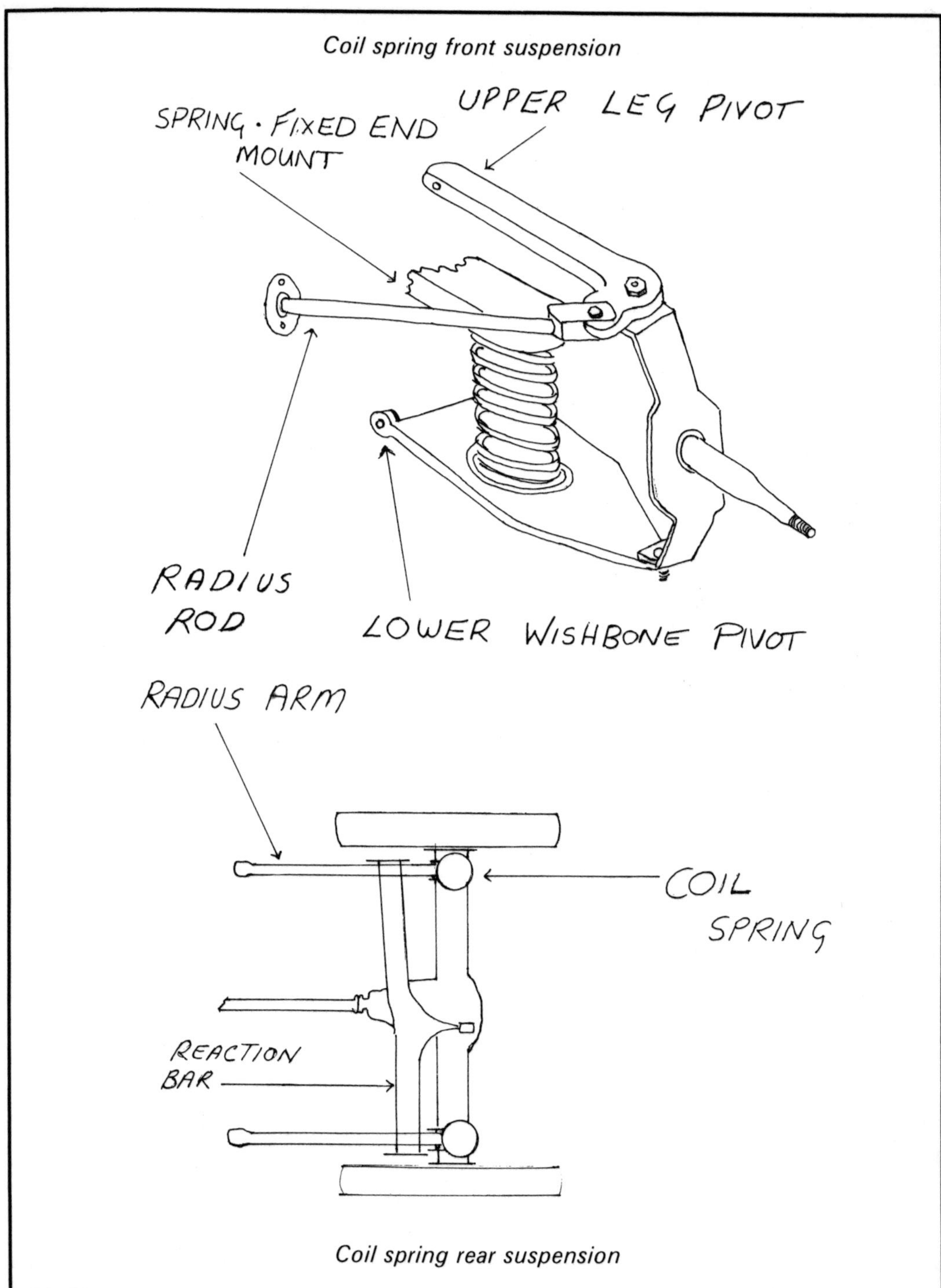

Coil spring rear suspension

The springs can be cut with a hacksaw or an electric angle grinder with a metal cutting disc. Springs from opposite sides of the car must be cut to the same length; where spring packing plates are located in the spring carriers these must be retained. On some designs of coil spring carrier it is possible to lower the vehicle ride height without cutting the spring. The illustration shows a spacer inserted between a lower wishbone and the spring carrier.

In lowering the vehicle by shortening the springs some of the suspension travel (up and down) will be lost. It is important to ensure that items such as 'bump stop' rubbers still leave enough travel; and the rubbers should be trimmed if necessary. The vehicle bodywork will also be lowered in relation to the wheels, so the clearance should be checked by fitting a wheel to the suspension, without the spring and carrier, and the suspension moved through its full range of travel.

Leaf springs. These are simple, strong, and can be used to locate the axle they provide suspension for. Leaf sprung cars can be lowered by two methods: where the axle is mounted above the spring, by fitting lowering blocks between the axle and the spring; or by having the spring re-tempered to a different profile.

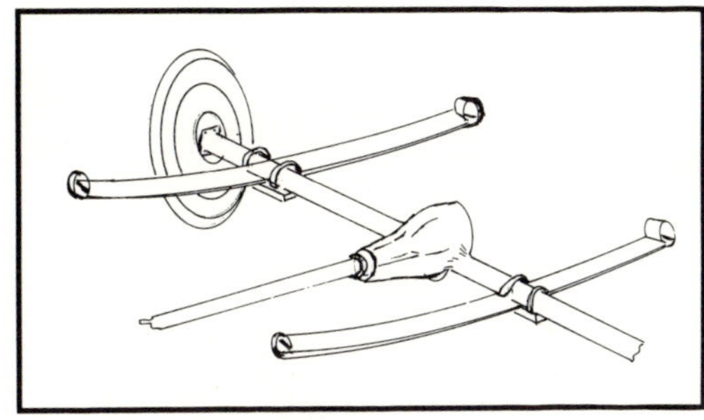

A leaf spring rear axle

On vehicles which use only the spring to locate the axle, axle 'tramp' and poor handling can occur when under compression loading the spring assumes some degree of 'S' curve; this shortens the distance between the axle centre and the spring eye, causing the axle to skew.

Blocks to lower a leaf spring axle

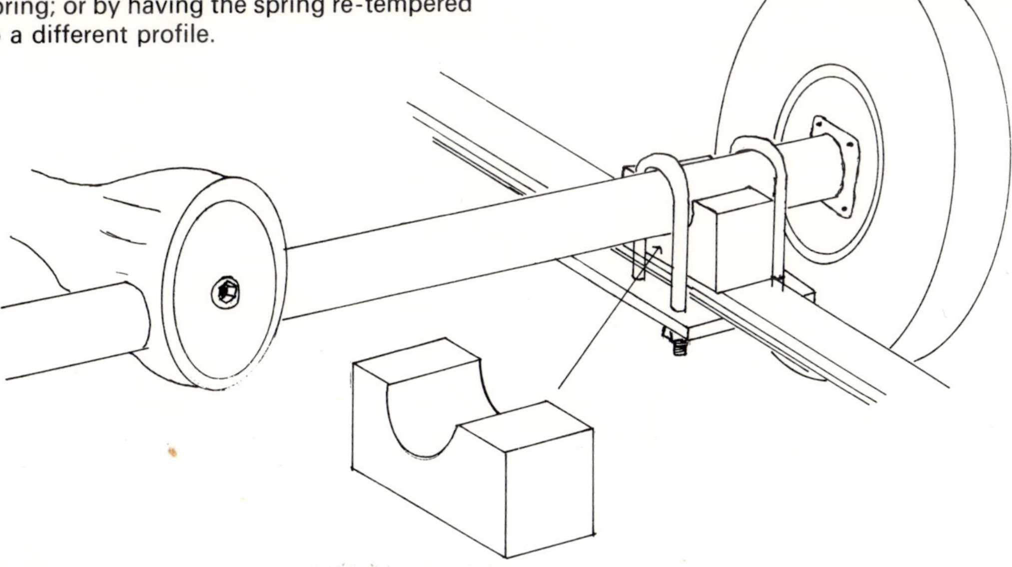

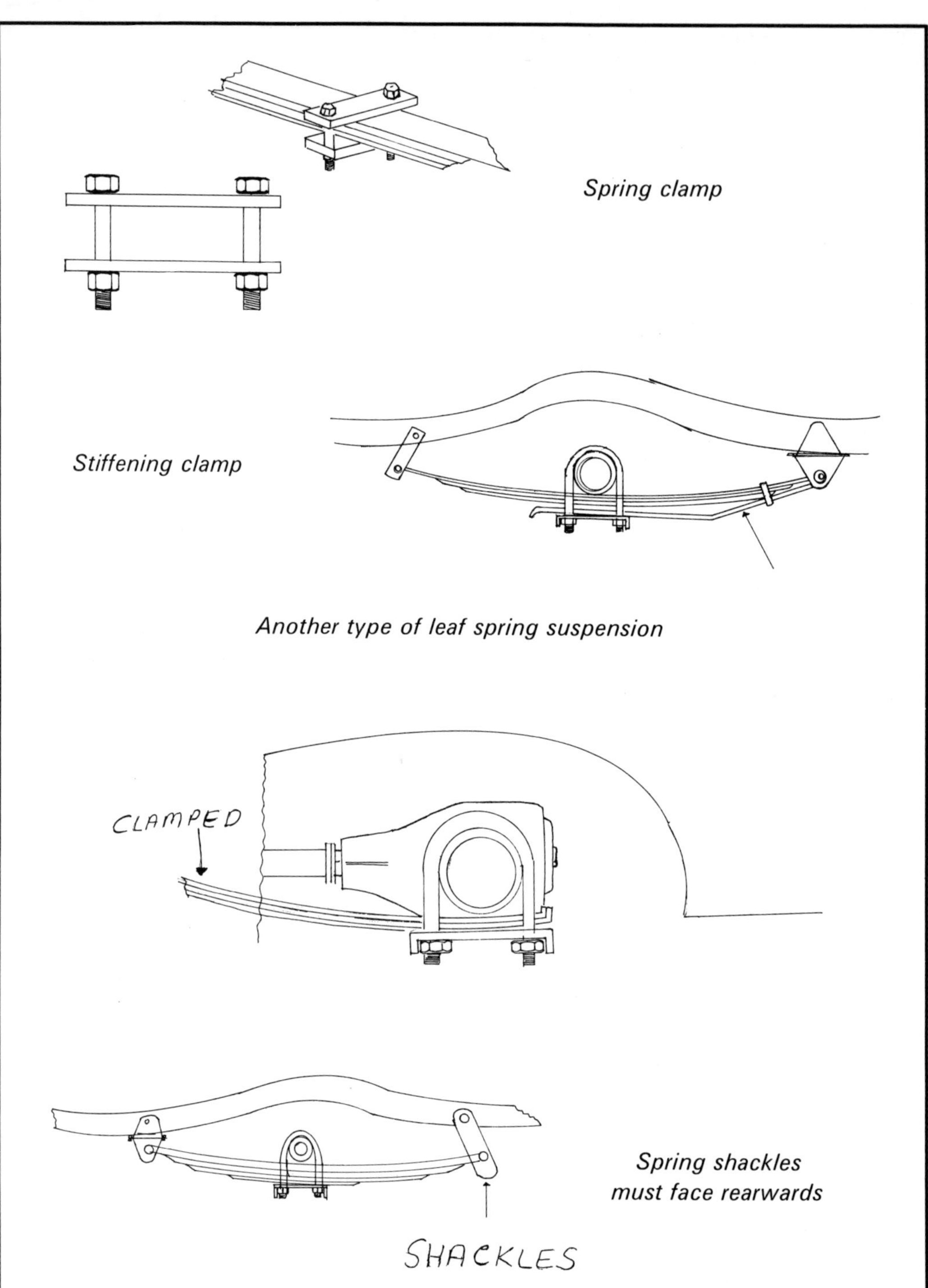

Spring clamp
Stiffening clamp
Another type of leaf spring suspension
CLAMPED
SHACKLES
Spring shackles
must face rearwards

The length of spring behind the axle does most of the spring work whilst the forward portion does most of the locating. Ensuring that the forward portion of the spring is firmly clamped together will make that portion a more effective 'trailing arm'. The spring should be very firmly clamped together with either spring clamps made from stout bar stock bolted together with high tensile bolts, or a blacksmith can forge straps all around, welding up the overlapping ends.

Another method of stiffening the forward portion of a leaf spring is to fit a device, as shown in the illustration, that bolts to the spring saddle and clamps the main spring behind the spring eye.

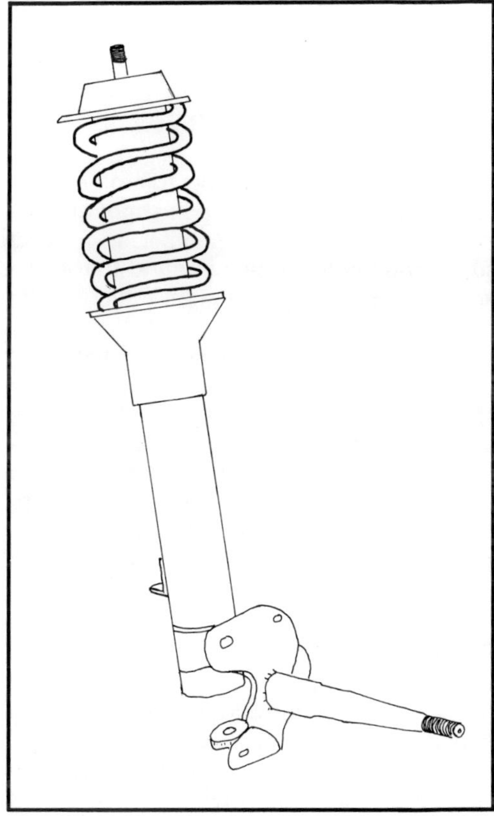

The Macpherson strut

The MacPherson strut. This device is a combination of suspension, shock-absorber, and wheel location. The strut is formed as a leg, with the shock absorber as an insert and a coil spring as suspension.

The MacPherson strut cannot be modified without the proper tools and equipment, but some improvement can be made by fitting competition strut shock-absorber inserts. Complete modified competition strut units are available for most vehicles.

Mini rubber suspension. Only the rubber suspension units fitted to the Mini range are suitable for racing use; the Hydrolastic units cannot be modified and are not suitable for competition work.

To lower the Mini suspension the rubber spring unit, together with the trumpet on the end, should be removed (see the relevant Haynes workshop manual for exact details). The attachment knuckle is then driven out of the rubber spring unit and about $1/8$in of rubber cut off.

The four rubber spring units should be cut down by an equal amount, refitted to the car, and the actual amount the suspension has been lowered measured. The usual amount by which a racing Mini is lowered is about 2 inches. Short racing shock-absorbers should then be fitted.

Whilst lowering a Mini suspension the wheel camber angle should also be altered to about $1^1/2$ degrees of negative camber. On the rear suspension this is achieved by removing the outer brackets which retain the suspension spindles, welding up the existing hole, and redrilling one $1/4$in higher. On the front suspension the camber is altered either by cutting and extending the bottom suspension arm by $1/4$in, or replacing the arm with a commercially made rose-jointed arm. Cutting and extending a suspension arm should be done by a professional welding shop – the arm is subject to considerable stress, especially in racing use, and its strength when welded must be guaranteed.

Torsion bars. Torsion bar suspension is simply lowered by adjusting the bar carrier.

Stiffer spring rate torsion bars are available; but are expensive.

One solution to increasing the spring rate on a torsion bar vehicle is to fit additional springs together with, and over, competition shock-absorbers. Many torsion bar sprung cars have sufficient clearance around the shock-absorber mounting area to do this.

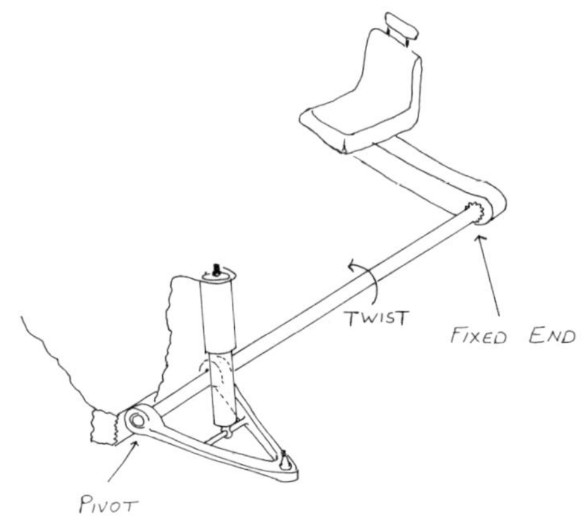

Torsion bar

Suspension location

The illustration shows some of the suspension location devices you may encounter. The workshop manual for your car will give you the factory dimensions for these components.

On most of the cars suitable for low budget racing none of these components are adjustable: the settings are determined by the length and dimensions of the component. On dismantling and rebuilding the suspension you should check that the actual dimensions of the components agree with the dimensions given in the workshop manual. Accidents and 'kerb bumps' can damage and alter the dimensions of such components.

If required, some modifications can be made to a non-adjustable suspension that will give some small range of adjustment.

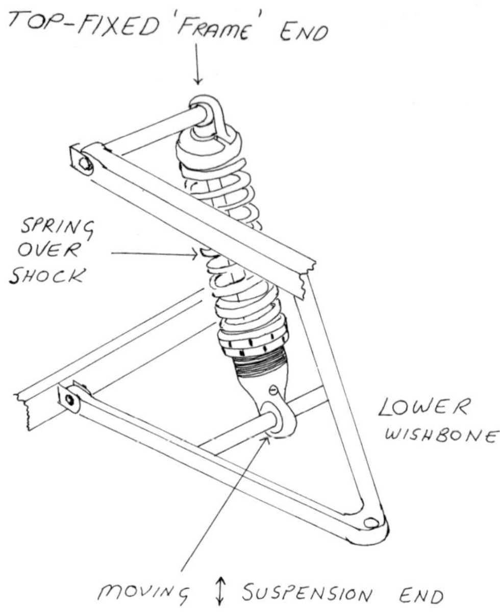

Coil spring over shock absorber

The illustration shows the top of a front suspension strut with an adjustment modification for variation of camber. In making such modifications bear in mind that small adjustments can have a great effect on the handling of the car.

To eliminate 'compliance' in a ball joint or bonded-rubber bush, a rod end or spherical bearing can be substituted. These are precision-made devices and are available in various grades depending on their intended task. The rod ends are used to replace track or radius rod ball-joints; the spherical bearings are for location in castings and pressings.

Steering gear

Excess movement or wear in the steering box or components should be adjusted as explained in the vehicle workshop manual.

The steering 'feel' can be altered by substituting a steering wheel of a different diameter. Note that the smaller the steering wheel the heavier the steering will become.

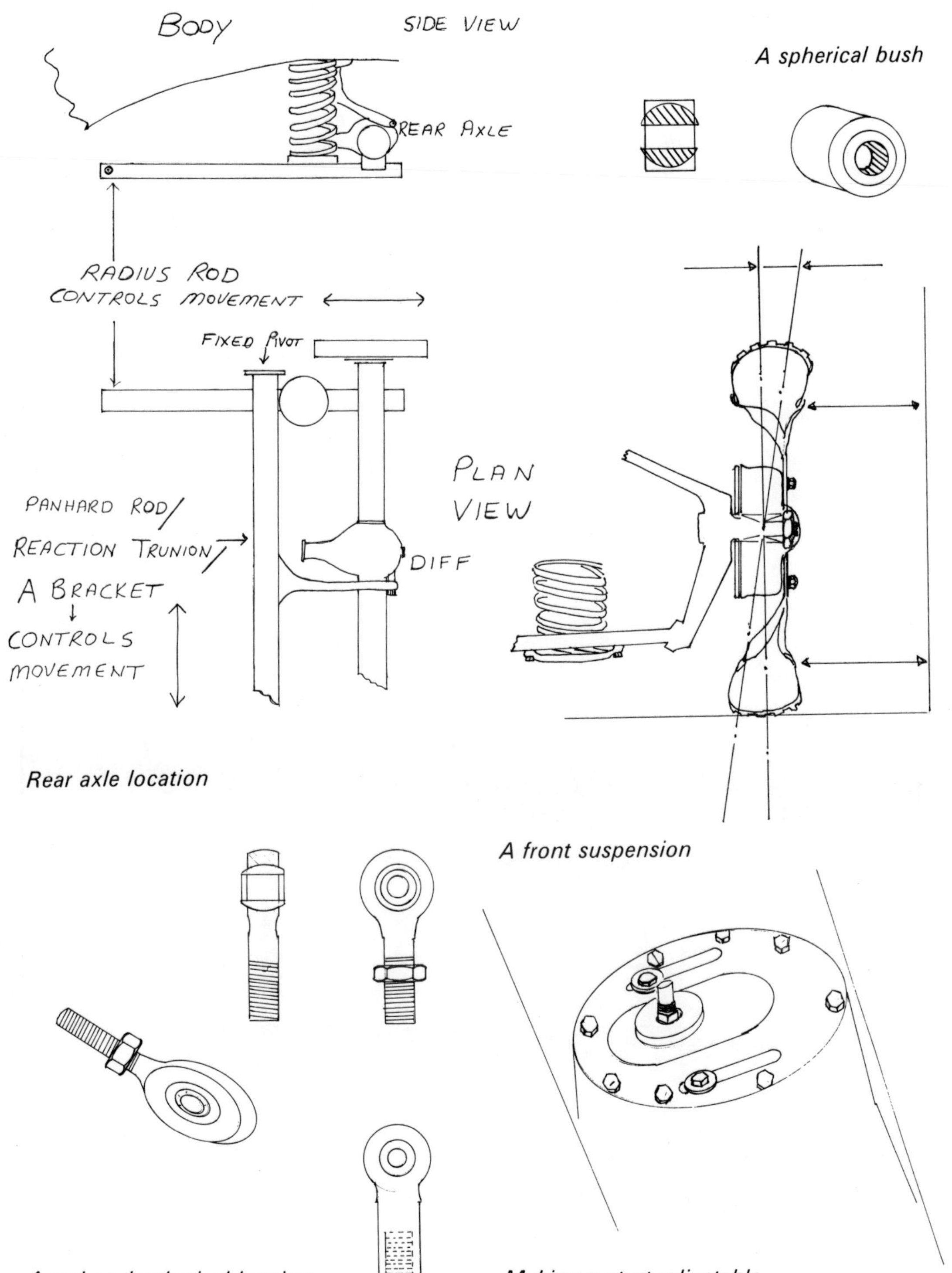

Rear axle location

A spherical bush

A front suspension

A rod-end spherical bearing

Making a strut adjustable

Shock-absorbers

Racing shock-absorbers are much firmer in action than the units usually supplied as original equipment on most cars. Shock-absorbers cannot be modified by the amateur mechanic and the correct type for the model of car should be purchased from a specialist racing parts dealer.

Most racing shock-absorbers can be rebuilt by the manufacturer, and second-hand units, if reasonably priced, are a worthwhile purchase. Some racing shock-absorbers are simply a firmer version of the standard item, but the best are adjustable either for the total damping effort, or for bump and rebound separately. Adjustable shock-absorbers allow tuning of the suspension peformance to suit varying tracks and conditions.

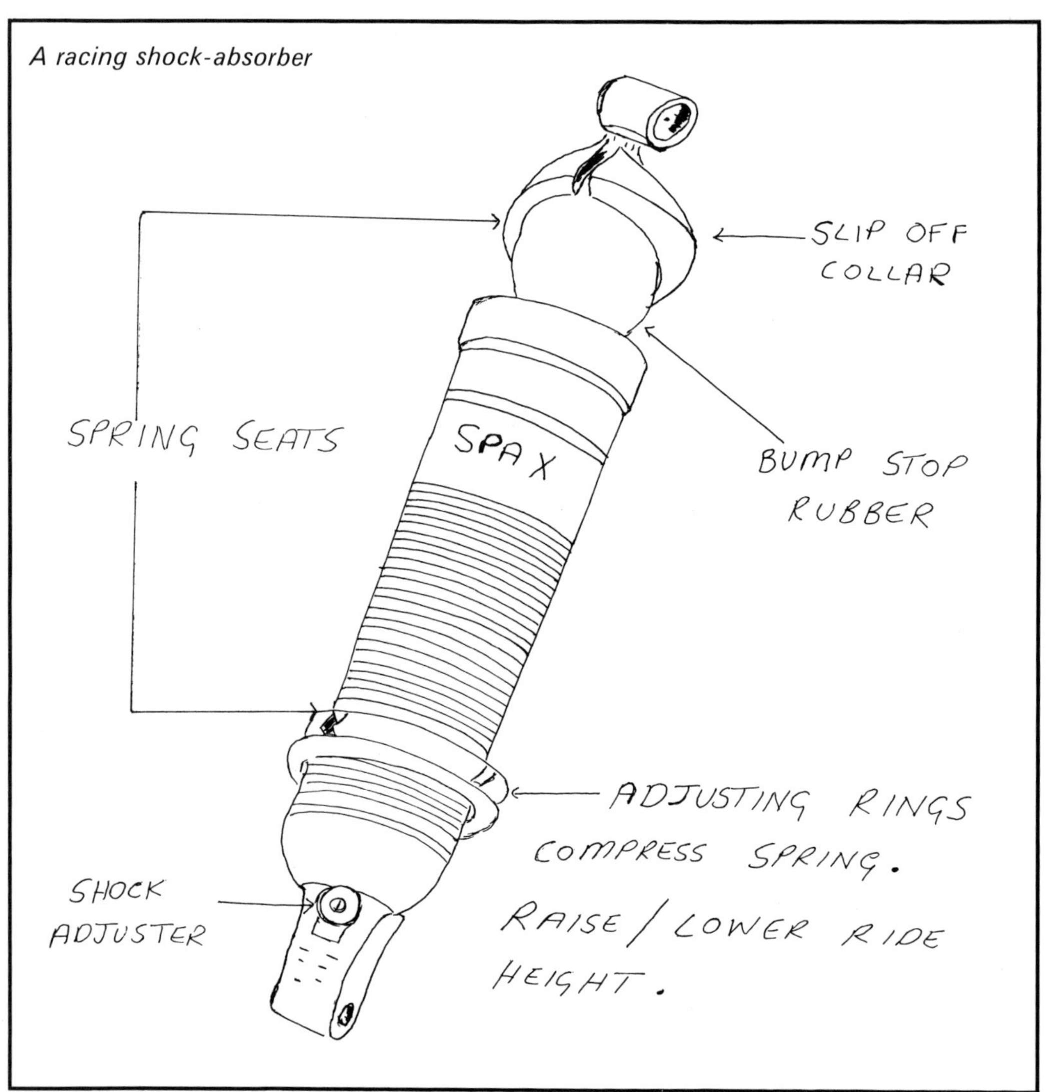

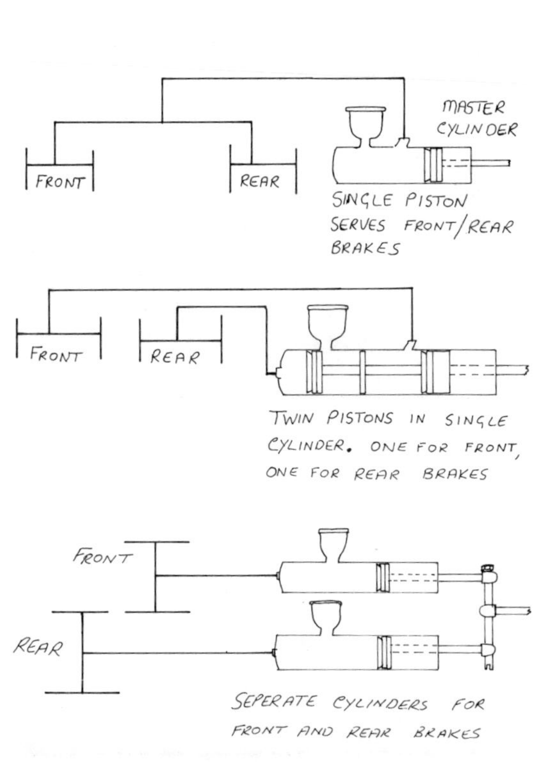

Single and twin piston brake master cylinder

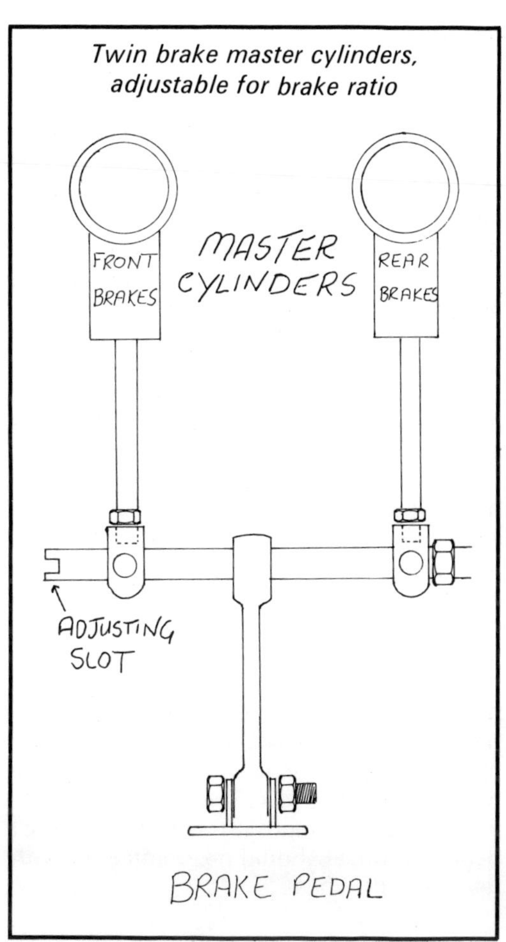

Brakes

With low budget amateur racing cars the brakes fitted to the vehicle as original equipment are perfectly adequate, provided that they are in first-class condition. A thorough brake overhaul should be carried out and any faulty parts renewed or refurbished. Flexible brake hoses should be renewed on rebuilding and then inspected regularly. They can wear quickly if subject to chaffing and are easily damaged by stone chips. A worthwhile modification, although expensive, is to replace the hoses with 'Aeroquip' stainless steel overbraided hose which improves brake 'feel'.

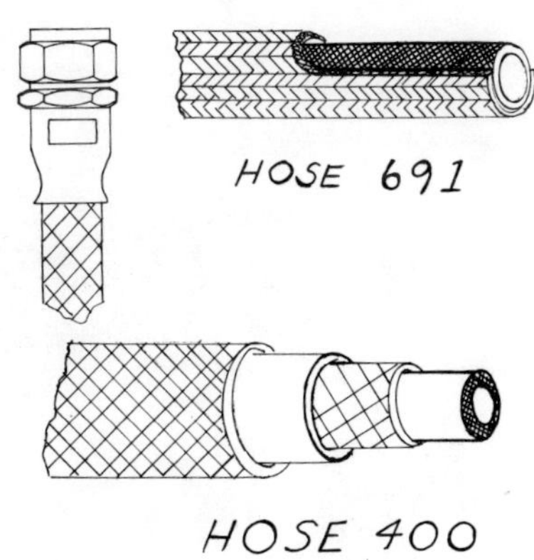

'Aeroquip' brake hose

Wheels and tyres

The tyres should be removed from the wheels and the wheels carefully examined for evidence of splits or cracks, particularly around the welded areas. The wheels should be thoroughly cleaned back to bare metal with a wire brush and emery cloth, or by grit blasting; then painted with a thin coat of aluminium paint – any cracks that develop can then be seen quite easily.

Fitting wheels and tyres of a wider dimension than the original equipment is a popular method thought to improve a car's roadholding. This is not always the case and many suspension systems are designed around the wheel width and tyre profile originally fitted: often, at best the suspension cannot utilise, for instance a low profile tyre, and at worst may be adversely affected. Fitting much wider than original wheels may also seriously overload the stub axles and wheel bearings.

Slightly wider wheels and tyres can, however, be of some benefit, and companies such as Wella Wheels make a range of plain steel wheels at very reasonable prices to fit most cars. The wheel 'offset' is an important dimension and the original degree of offset should be maintained with replacement wheels.

The RAC Tyre Regulations or club and race class regulations dictate which type of tyres a particular car must use. In most amateur racing classes the choice is very wide, and suitable cheap tyres can often be purchased from advertisers in *Motoring News*.

Specialist racing tyres, such as the slick (treadless) tyre, are not used in amateur motor racing apart from on the highly modified cars. When used on such cars, the slick tyres are usually purchased second-hand from professional Formula Ford 2000 or Formula Three teams. Slick tyres used in

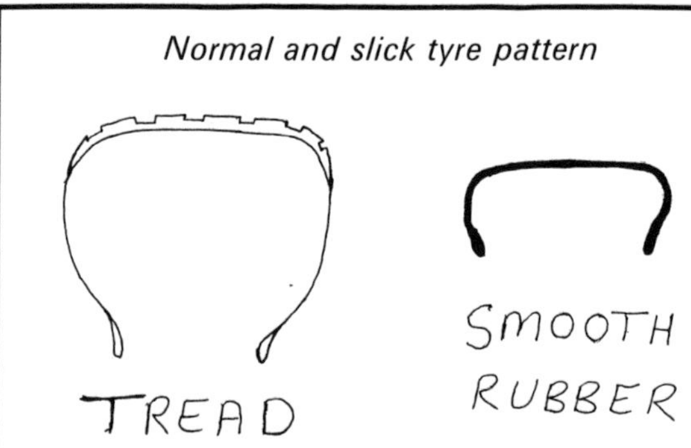

A highly modified Capri on slick tyres

amateur motor racing will often last a whole race season, whereas the professional single-seater formulas use a new set of tyres for each race.

Example suspension modifications

Vehicle: Alfa Romeo GT Junior.
Owner: The author.
The Alfa Romeo GT Junior is fitted with coil springs for the front and rear suspension, the front and rear pairs being of different dimensions.

The front springs are removed by separating the bottom suspension wishbone from the top arm and radius rod by undoing the swivel nuts of the stub axle. The front springs are under a considerable pressure and must be held compressed during removal using the correct special tool. Do not try and jury-rig a tool – hire the right one from your local hire shop. The spring carrier in the lower wishbone unit will be fitted with a shaped spring end plate and, sometimes, shims used to set the vehicle ride height exactly (side to side/front to rear). Any shims present must be retained and reused in their original spring

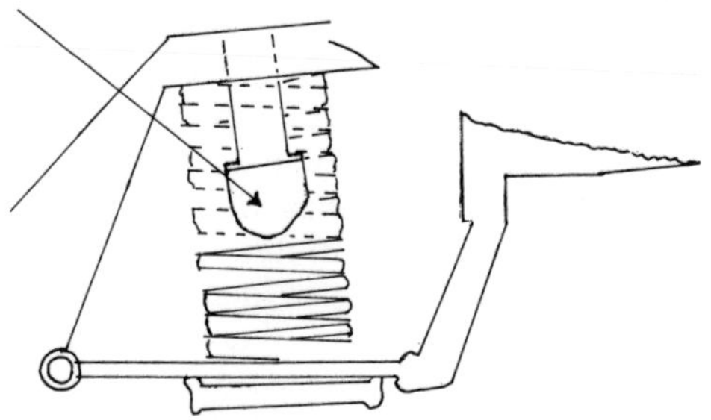

The bump stop rubber

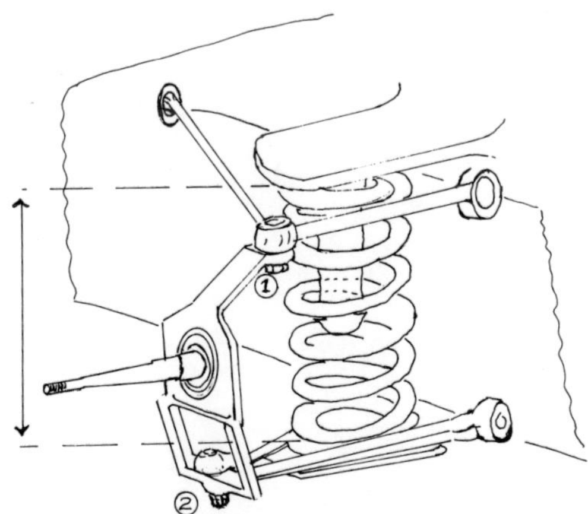

Some coil springs are under considerable compression between points 1 and 2

carrier. Whilst the front suspension is dismantled the bump-stop rubber cones, mounted inside the coil spring, must have some rubber cut off their length; this is so that sufficient suspension travel remains with the shortened springs.

The live axle rear suspension is simple to dismantle as detailed in the model workshop manual. All the springs are shortened by one coil. Note that all the springs must be shortened by exactly the same dimension.

To complete these basic suspension modifications the car was fitted with four Spax adjustable competition shock absorbers.

Considerable further suspension modifications are possible, and allowed, on cars intended for entry in the fully-modified race class. The radius arms, at front and rear, can be fitted with spherical bearing rod ends; as can the anti-roll bar ends. The front anti-roll bar is attached to the body/chassis in two rubber cushions; these can be replaced with solid mounts machined from alloy or other suitable material. For further roll control a rear anti-roll bar can be fitted.

In order to use slick type racing tyres on wider wheel rims, the suspension geometry can also be altered.

Vehicle: Austin A35 (first reg 1956)
Owner: Andy Anderson, Wood Green, London.
Driver: Nick Amey, Knebworth, Warwickshire.
Team manager. Wendy Markey. London.
Winners of the 1984 Classic Saloon Car Championship. Car owner and builder Andy Anderson purchased the car as a rolling-chassis, after it had already been used for some motor racing; he also purchased the remains of another racing A35 after it had been written off in an accident. Andy's car was rebuilt from the two. The steering had been converted to a rack and pinion type, which did not conform to the Classic Saloons race class regulations. Whilst rebuilding the car the steering was reconverted back to the original type (worm and roller). The front springs (coils) were shortened heavy duty units. The rear suspension was lowered by having the springs (leaf type) professionally reprofiled and tempered to a flatter profile. Because the car was to be fitted with a quite highly tuned engine, Andy fitted specially toughened half-shafts to the rear axle whilst working on the suspension. The original drum brakes fitted to both front and rear were converted to disc brakes all round, by using parts from a scrap Austin-Healey Sprite.

Andy completed the suspension modifications by fitting four adjustable racing shock-absorbers.

Vehicle: Ford Capri
Modifications to suit entering a car in a 'single circuit' series of a road-going race class.

Road-going race class series aim to keep the cost of entering a competition down by banning all but the most minor modifications, such as might be carried out by any enthusiastic sports motorist. A Ford Capri or similar car can be made quite suitable by simply fitting competition front shock-absorber strut inserts and competition rear shock-absorbers. A further small handling improvement can also be gained by fitting a stiffer anti-roll bar.

5

The engine

MOTOR RACING produces greater stresses within an engine than even the most harsh treatment in normal road use. The engine should be completely stripped down and rebuilt: any parts of suspect condition must be replaced with new components. The RAC Blue Book gives concise details of any engine tuning work that is allowed for a particular class of racing car.

There are three ways by which you can obtain an engine in suitable condition to use as a racing engine:

1 By purchasing a complete secondhand racing engine. With this method you will have to consider the cost with respect to the total that you have to spend on the car; and be sure that the engine is in good condition and has some form of guarantee.

2 By having an engine rebuilt by a professional engine builder. The cost to rebuild an ordinary BMC or Ford engine would be between £250 and £350, plus the cost of some specialized parts, such as a camshaft more suitable for racing use. With a rebuilt engine you should receive some guarantee in respect of its performance.

The additional cost of this method means that on a restricted budget you will have less money to spend on the car and its general rebuild. By initially selecting a car that is in good condition you should still be able to complete the project within a £1000 budget.

3 By rebuilding the engine yourself. Many engines contain fewer than sixty moving parts, and using the workshop manual for your car, a rebuild should not prove to be a difficult job.

Constraints on engine performance

Most engines in normal road cars are capable of producing more power than they do. The cost of extracting more power from an engine manufactured in a particular state of tune is: an increased wear rate, and an increased cost of production.

The tolerances used for the mass production of engine parts results in these components being out of balance at high revolutions per minute (rpm), whilst the cost of some materials or manufacturing methods also means that some components are unsuitable for high rpm use. When rebuilding an engine for racing use these constraints must be overcome: by using more suitable materials for some components, balancing others, and accepting a higher rate of wear for all the components.

Rebuilding the engine

Vehicles suitable for amateur motor racing

on a low budget will be fitted with either: a pushrod overhead valve engine, or an overhead camshaft engine.

Before undertaking any work on the engine you must check what is allowed in the race class regulations. Some racing classes require the car's engine to be in the standard roadgoing configuration, whilst others allow various degrees of tuning.

Tuning one aspect of an engine, such as fitting a higher lift camshaft, often drastically effects other components. Apart from minor modifications, such as fitting bigger carburettors or a free-flow exhaust, attention to one item will result in it being necessary to make many alterations right through the chain of its effects.

The workshop manual for your choice of car will clearly detail the step-by-step stripping and rebuilding of the engine. By following the instructions in the workshop manual, together with the following list of points that relate to racing use, you should be able to rebuild the engine to a suitable condition for amateur racing use.

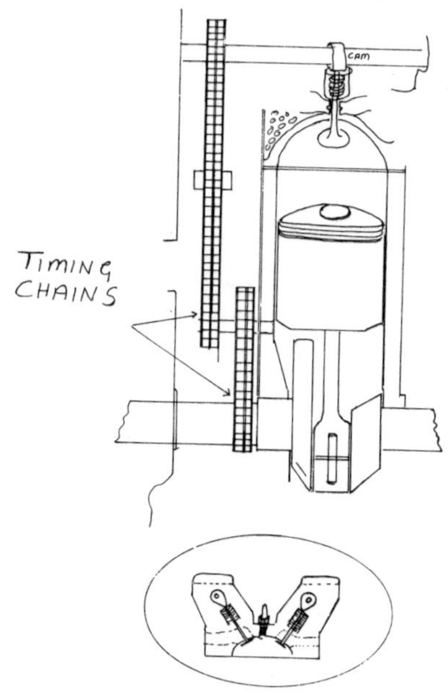

Overhead camshaft engine

Pushrod engine

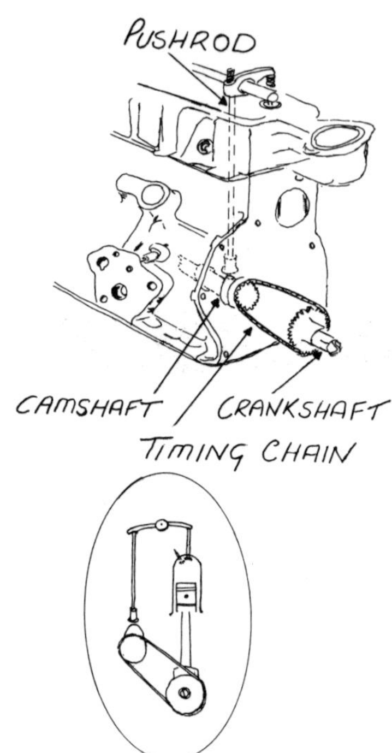

The cylinder head

The cylinder head must be in good condition, with a perfect gas-tight seal of the valves. Where gas flowing and polishing is allowed in the regulations, this can be carried out using small grinding stones and flap wheels attached to a flexible shaft and electric drill.

Modified cylinder heads, with larger valves and other features, can often be purchased secondhand. On older engines with quite a low compression ratio, it is possible to have a small amount of material ground off the face of the head to raise the ratio. If the compression ratio is raised it is important to ensure that the cylinder head bolts do not 'bottom' in the thread sockets, otherwise the cylinder head cannot be tightened down to the correct torque setting. The bolts should be shortened by an appropriate amount.

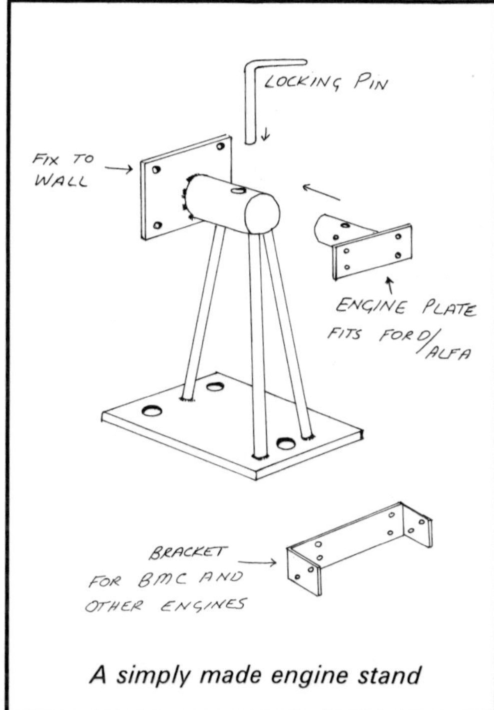

A simply made engine stand

Valve springs

Stronger than original equipment valve springs are essential for racing use. Valve 'bounce' often takes place well before the maximum rpm limit that can be extracted from a roadgoing engine. Stronger valve springs do place a higher loading on the rest of the valve gear so camshaft(s) pushrods and tappets must be in good condition: all these components will be subject to higher wear when using stronger springs and should be regularly inspected.

Camshafts

Where the regulations allow, an improved profile camshaft can be fitted. Such camshafts give an increase in performance at the sacrifice of some other characteristic, such as low speed tractability. Fitting a higher lift camshaft results in the valves extending further into the combustion chamber when fully open, with the possibility that the valves will contact the

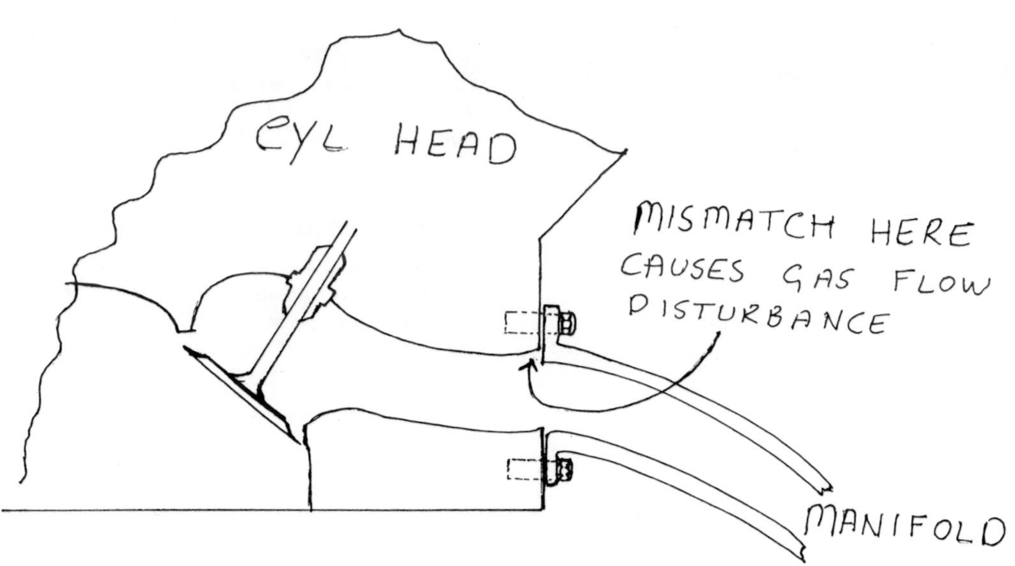

Some class regulations allow grinding to flow-match the manifolds to the cylinder head

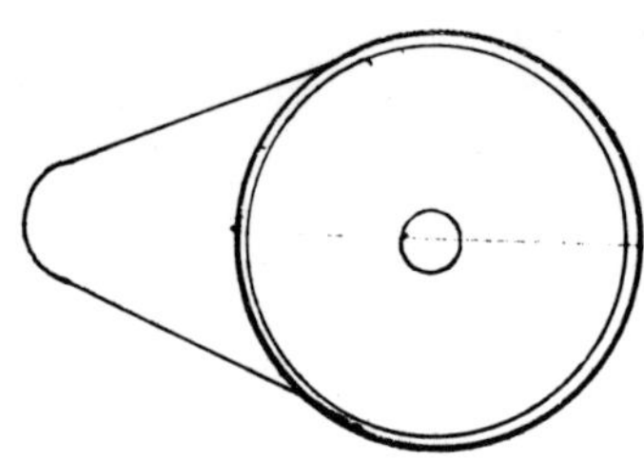

The camshaft 'profile'

piston. Where this is the case the pistons should have a 'pocket' machined into them to allow for valve clearance. Rally and speed shops which supply performance camshafts will often carry out just the cam fitting and modifications for you, leaving you to complete the engine rebuild.

Note that valve timing often changes when fitting an 'improved' camshaft: if using the original timing wheels, the timing marks on them may no longer be valid. New timing data is always supplied with improved camshafts.

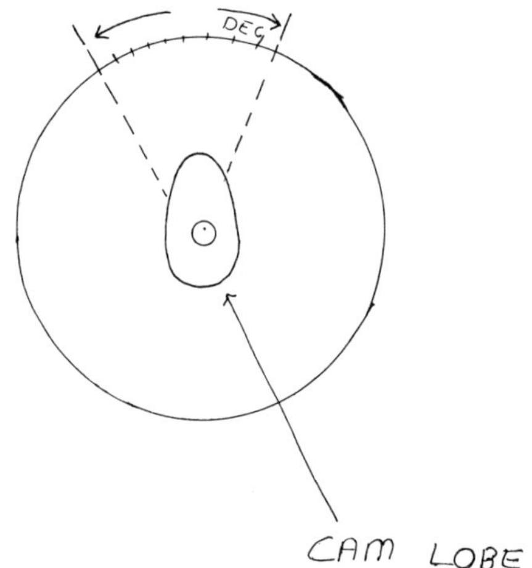

The camshaft 'period'

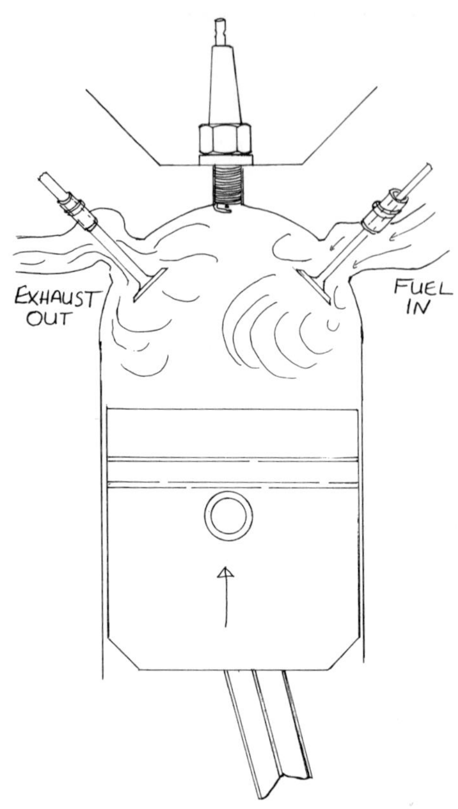

Valve overlap period, when the flow of gases in to the head helps the exhaust flow out

Timing chains or gears

A new timing chain or gear should always be fitted. It is impossible to ensure accurate camshaft timing with a chain or gear that has any degree of wear in it. Many engines in the Ford and BMC range have a single camshaft timing chain. Duplex (twin) timing chain conversion kits are available for most of these engines and give improved strength and precision to the timing assembly.

Pistons and rings

A slightly increased bore clearance is desirable in a racing engine; this allows for additional piston expansion at high rpm and engine temperatures. If the engine is to be rebored this tolerance can be allowed for by the engineering shop carrying out the work.

Where a rebore is not required it is important to ensure that the 'glaze' on the cylinder walls is broken and roughed to allow for oil retension. This can be simply done with an emery 'glaze buster' wheel mounted in an electric drill.

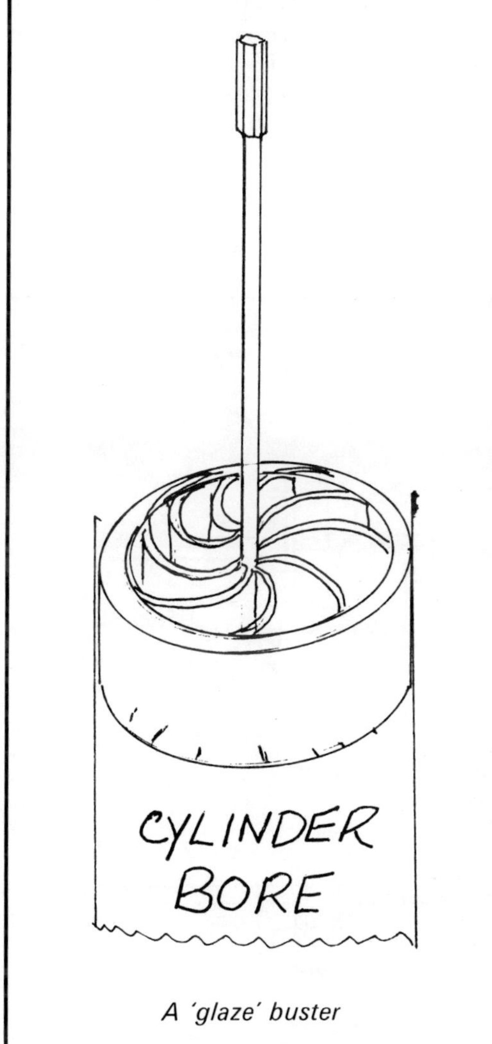

A 'glaze' buster

A cylinder hone

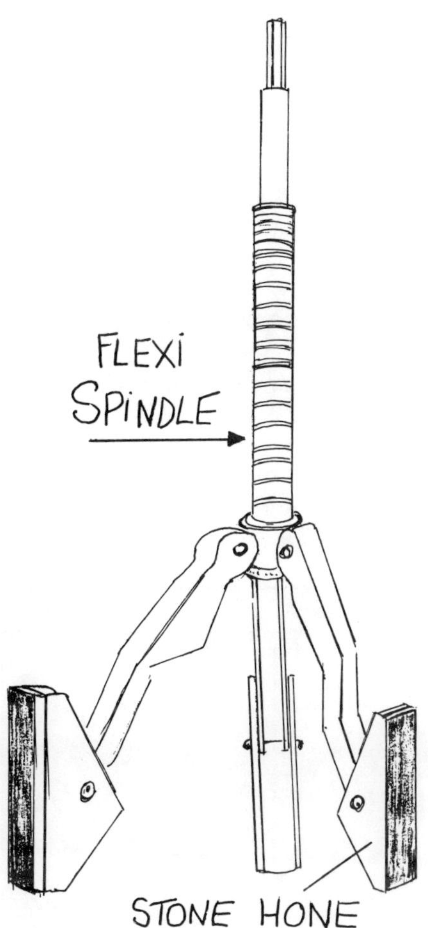

Crankshaft and bearings

When rebuilding these components, the whole assembly, including the connecting rods and pistons, should turn freely by hand. If a bearing is found to be tight, the bearing cap can be gently polished with oiled emery paper to ensure a free running fit.

Competition main and big-end bearing sets are available for many engines and cost very little more than the standard items.

Oil pump and sump

A new oil pump should be fitted and if a high pressure pump is available for the model of engine one should be substituted for the standard unit.

The sump pan of some engines have very few baffles to prevent oil surge. In racing, where centrifugal force during cornering can drag oil up the side of the sump pan away from the oil pump pick up point, welding additional perforated baffles in the sump pan can help to prevent this effect.

Dry sump equipment

A dry sump arrangement has a separate tank for the engine oil supply. A special dry sump oil pump is required together with a special shallow sump pan. Most amateur racing class regulations prohibit the use of dry sump oil supply, unless it was fitted to the model of car as original equipment.

Engine breather pipes

All engine breather pipes on a racing engine must vent into a catch tank. Where the engine has more than one vent pipe it is important to ensure that as much oil as possible can be returned to the sump. If all the vents are taken to a catch tank in individual runs it is possible for the engine to pump out too much oil at high rpm, leading to oil starvation.

Component balancing

Many mass production engines are not manufactured with the flywheel/crankshaft assembly as a fully balanced unit. At high rpm and stress this produces vibration and wear, and limits maximum rpm and acceleration.

The major turning and reciprocating components of the engine – flywheel and

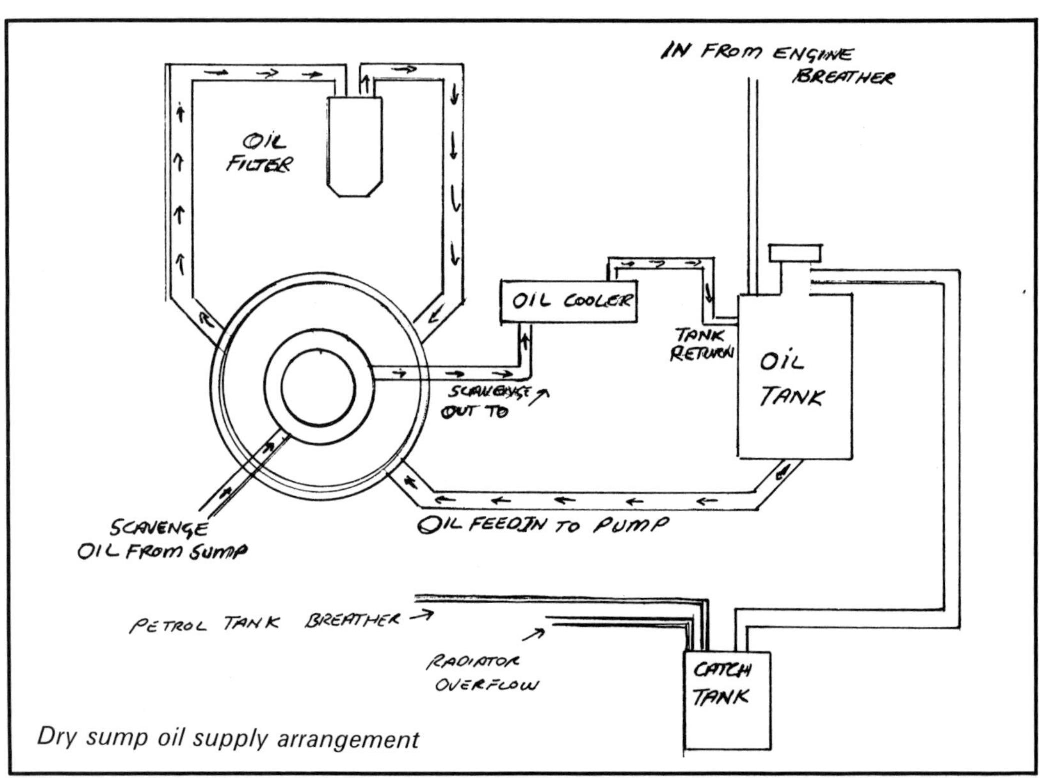

Dry sump oil supply arrangement

clutch, crankshaft, connecting rods and pistons – should be balanced as a unit. Specialised equipment is required to balance the crankshaft assembly and this work should be entrusted to an engineering company. A number of companies advertise in *Motoring News* and *Autosport* magazines: a typical cost to balance the complete assembly is about £35.00.

Carburation and exhaust

The carburettors should be fully refurbished. Reconditioning kits are available for most models. Worn carburettors make accurate engine tuning impossible as they will run rich one minute and weak the next.

In racing catagories and classes in which the standard roadgoing components are mandatory, improvements can still be made. The carburettor can be fitted with larger jets and a carburettor specialist can supply all the parts and instructions for fitting them.

Free flow air filters

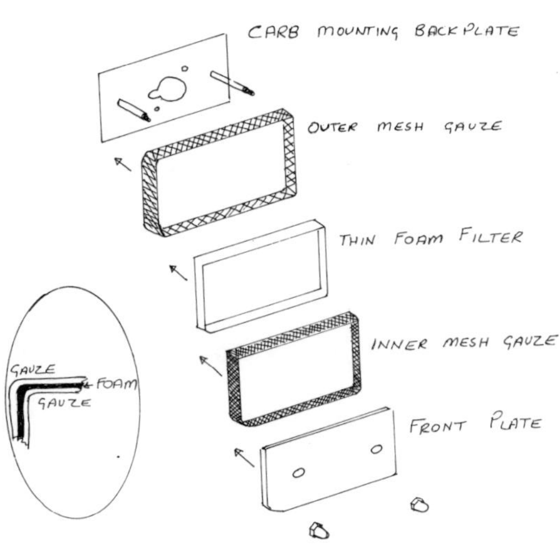

A standard exhaust system can be modified

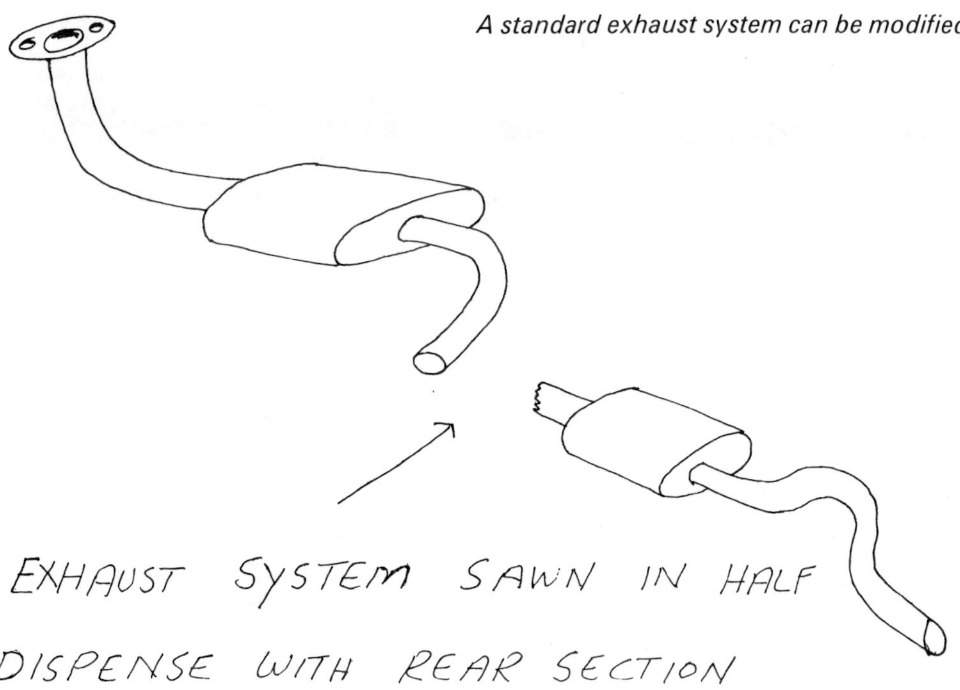

Where the racing regulations state that carburettors are 'free', that means that any modification is allowed. A better manifold, perhaps with twin carburettors, can be fitted. In the spares and parts section *Motoring News* has these types of item secondhand; the unit required can often be obtained at a reasonable price. The scrapyard can also be a source of parts; from a performance model of the same vehicle.

For racing use, the air cleaner boxes can be replaced with free-flow type filters.

Environmental pressures have reached the race world and virtually all race cars have to be silenced. At every race meeting at specific circuits, and at some time during the season at all circuits, RAC scrutineers will be taking noise (decibel) readings of the engine exhaust when running at about three-quarters throttle. The allowed reading for different classes is shown in the RAC Blue Book.

For many racing categories the standard roadgoing exhaust system is required. Where modification is allowed the accompanying illustration shows a useful simple change, which consists of shortening the system and bringing the outlet out at the side of the car. Where the car was equipped previously with two silencer boxes, one will often do, without raising the decibel level too much. There are many improved performance exhaust systems on the market and one should certainly be used if the budget allows.

Distributor and spark plugs

The distributor must be in good condition. You cannot set accurate ignition timing with a worn distributor, and without that nothing else can be set.

The vacuum advance section, if fitted, can be blanked off. Non-vacuum advance units for BMC or Ford engines can be obtained quite easily. The ignition advance at high rpm is the important setting for racing use. When finally tuning the completed car a 'rolling road' tuning

specialist can accurately set the ignition timing and carburettor settings at the optimum for racing use.

The ignition leads should be replaced with the straightforward wire type leads, but suppressed plug caps should still be used.

Spark plugs. Use the recommended grade for your engine to start and warm it up. For competition use change these for a set of harder (cooler running) plugs. NGK make an excellent range of plugs with grades BP6ES to BP9ES covering most models of engine.

The gearbox and differential

Many of the amateur racing classes require to be used the original equipment gearbox for that model of car. The workshop manual for your car will give details of gearbox renovation.

Reconditioned gearboxes for popular cars can be purchased very reasonably on an exchange basis. Often this price is less than that which it would cost to buy the parts to fix a gearbox yourself.

For some gearboxes, especially as fitted to Minis and the series that fits AH Sprites, A35s, Minors, and some Ford boxes, close ratio gear sets are available. The spacing between the gear ratios is narrower than for road use and this aids acceleration. The spares section of *Motoring News* is a good source for secondhand close ratio gear sets.

The differential should be in reasonable condition, but some wear can be tolerated. The final drive ratio of some cars can be altered by fitting a higher, or lower, ratio crown wheel and pinion from another model of the same manufacturer's vehicles: the same axle, but with different ratios, is often fitted to a range of vehicles.

Example engine rebuilds

It should be appreciated that only minor tuning work is required to build an engine suitable for low budget amateur motor racing. Serious engine tuning adds greatly to the expense of building and maintaining a car and many of the amateur racing formulas prohibit tuning in order to limit the expense involved in competing.

For the reader wishing to delve further into engine modifications there are numerous books available devoted entirely to tuning for all the popular production engines used for motor racing.

The BMC 998cc engine

As fitted to the Austin A35 and Morris Minor.

The 998cc BMC engine is a four-cylinder pushrod OHV engine of straightforward construction. These engines were originally manufactured in a large range of states of tune for fitment to other models from the same maker, such as the Austin-Healey Sprite and MG Midget. If the class regulations allow, a mildly-tuned engine, more reliable than a standard unit in regular competition use, can be assembled quite simply by rebuilding the unit using:

1. A part number 12G202 cylinder head as fitted to manual transmission Austin and Morris 1100s: or, a part number 12G295 cylinder head (which has larger valves) as fitted to the 998cc Mini Coopers or MG 1100s.
2. Cutting exhaust valve clearance pockets in the cylinder block, if required by the valve size.
3. Fitting an improved profile camshaft.
4. Fitting a duplex timing chain.
5. Strengthening the centre main bearing cap.
6. Fitting heavy duty bearing shells.
7. Fitting teflon piston gudgeon pin caps.
8. Balancing the crankshaft, flywheel and clutch assembly, pistons and connecting rods.
9. Fitting an uprated oil pump.
10. Fitting a competition distributor.
11. Fitting a larger pulley to the generator to reduce its RPM in relation to engine RPM at high speeds. As detailed in the model's workshop manual, the engine should be completely stripped down, all the parts examined and a list made of those requiring replacement.

Cylinder head. Either of the cylinder heads listed will fit straight onto the 998cc block. The efficiency of the head can be improved by smoothing the surface of the combustion chambers and polishing them. Full cylinder head modificiations would include improving the combustion chamber shape and equalizing the volumes. The compression ratio can be raised by having a small amount of material machined off the cylinder head face: for mild tuning this would be about 0.050in.

Sometimes machining the cylinder head results in there being insufficient valve rocker adjustment; this can be regained by countersinking the tappet screw hole in the rocker arm and grinding away part of the top of the pushrod, so that it does not foul the rocker.

The cylinder head, with valves assembled to it, should be 'try-fitted' to the block, without a gasket, and the valve-to-block clearance checked with the valves fully open: the minimum clearance is $^1/_{16}$in. If there is insufficient clearance for the valves the block must have 'pockets' cut in its face for clearance. This can be done by an engineering shop or at home using an old valve with tool-steel cutters brazed to its face and using the cylinder head as a fixture for position. The cutter/valve is fitted to the cylinder head which is lightly bolted to the block, a tap holder can then be used on the valve stem to turn the valve and cut the pocket.

The cylinder head mating surface of the block should be checked for truth with a straight edge. On assembling valves and springs to the cylinder head, particularly stronger racing springs, the springs should be checked for coil-bind (when the spring coils are crushed up tight together) when the valves are fully compressed. With a mildly improved camshaft fitted, this should not be a problem; but high lift camshafts open the valve further. Coil-bind can be cured either by using valves with longer stems or by pocketing the spring base in the cylinder head: the latter is the usual method.

Camshaft. There is a large range of improved camshafts available to suit these engines. For mild tuning the higher lift type profiles should not be used as they would only over-stress the normal valve train components.

Rally shops and similar outlets are usually agents for reputable camshaft manufacturers and can advise on suitability and valve timing. Of those manufactured by the engines' makers the AEA 630 used in the 998cc Mini Cooper, or the AEG 510 used in the Cooper S are suitable.

Duplex timing chain kits are available from a number of suppliers and are a straightforward replacement for the original item.

Centre main bearing cap. This may be strengthened by machining the base of the cap flat, and backing it up with a suitable piece of bar steel stock: longer, high tensile, bearing cap bolts must be used.

Bearings. Heavy-duty bearing sets are available from a number of manufacturers for both main and big-end bearings.

Crankshaft. The crankshaft should be fitted to the block, and the bearings and the caps tightened to the correct torque: it should be possible to spin the crankshaft freely by hand. If it is not, the bearing caps should be loosened in turn and the tight bearing located: the bearing cap bearing seat and block bearing seat should be examined and cleaned and, in an extreme case, the seats can be very gently polished with oiled emery cloth to ease the bearing clearance.

The crankshaft, flywheel and clutch assembly, pistons and connecting rods, should be entrusted to a reputable engineering shop for balancing. Together with the balancing work it is advisable to have the connecting rods crack tested, checked for straightness and balanced end to end.

Teflon gudgeon pin inserts should be installed in the pistons when the engine is finally built up: these ensure that at high rpm the gudgeon pins cannot come into contact with the bores.

An uprated oil pump should be installed at this point and the engine built up according to the workshop manual details. Careful attention should be given to any change in camshaft timing made necessary by the installation of an improved-profile unit.

A non-vacuum competition distributor should be used if possible, or the vacuum advance unit sealed up; the ignition timing being set for near maximum rpm and finally adjusted on a rolling road.

A larger pulley should be fitted to the generator, which will reduce the generator rpm in relation to engine rpm, otherwise

Strap for main bearing cap

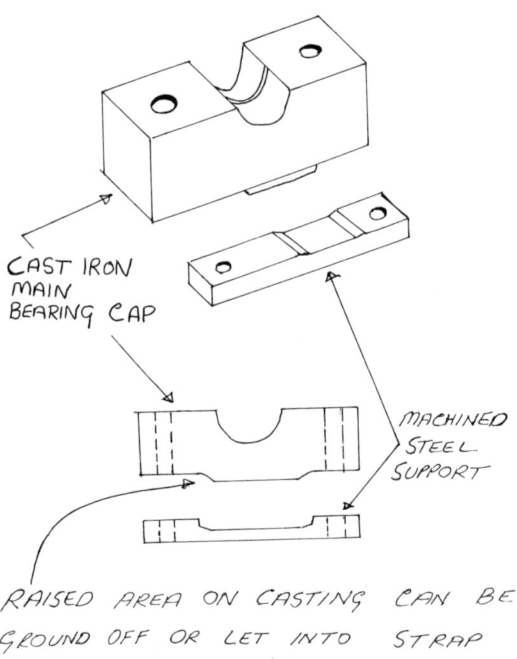

prolonged use at high speeds can lead to the armature disintegrating.

If the car is only to be used for racing the water pump-mounted radiator cooling fan can be removed: quite sufficient cooling is obtained with the car moving at speed and the fan merely absorbs engine power without benefit.

Carburation. There is a wide variety of carburettors and manifolds available for use on an improved 998cc engine; the type and class of races for which it is to be entered will influence the choice. A cheap improvement for general competition use can be made by using a secondhand twin carburettor unit from the AH Sprite/MG Midget range.

The Alfa Romeo twin-cam engines

These are four-cylinder twin overhead camshaft units, the basic design of which has remained unchanged for many years. The engine was manufactured, in sizes from 1300cc to 2 litres, as a high performance sports car engine and consequently is well tuned. The earlier units (late 1950s to late 1960s) can benefit from machining the cylinder head to raise the compression ratio, and the fitment of improved profile cams. These engines were manufactured as a balanced unit.

The twin overhead camshafts are seated directly in the cylinder head (without a separate carrier) and are driven by a chain from the crankshaft via an idler gear. The unit is quite straightforward to dismantle and rebuild, with the proviso that caution is observed in handling the many components which are made of alloy: scratches and nicks in the alloy can easily become failure and fracture points.

Great care should be taken in stripping the engine: the main bearing caps, for instance, are often difficult to remove but are easily damaged by the use of pliers or such like to remove them. Similarly the cylinder head can be difficult to remove because of reaction between the alloy and the steel cylinder head studs: with the engine stripped these studs should be thoroughly cleaned back to bare metal; any that are pitted or suspect should be renewed.

With the engine completely stripped as detailed in the workshop manual, the unit can be carefully examined for damage caused by corrosion of the alloy. The unit is fitted with steel 'wet' cylinder liners, and in an older engine these will probably be held fast in the block; a puller should be obtained or made to remove them. The base of the cylinder liners is water-sealed to the crankcase by rubber 'O' rings. Although the liner may seem firmly embedded in the block, once released from compression by removal of the cylinder head, they are likely to leak. The liners must be removed, the seat cleaned, and new 'O' rings fitted.

Provided the pistons and bores are not too worn they will be quite suitable for racing use. After a high mileage the bores will have a glazed finish which does not retain oil. With a unit of this type – with removable wet liners – a useful tip is to have the liners, pistons, and connecting rods vapour bead blasted. This process uses tiny glass beads which thoroughly clean the surface but does not damage it. On the steel bores it will give an excellent oil-retaining finish, whilst the alloy parts will be matt polished. Care should be taken to keep each piston with its matched bore, and new piston rings should be used.

On refitting the liners to the bores in the block care should be taken to observe the manufacturer's recommended tolerances; the liners are held in place by the cylinder head and leaving one liner more proud than the others can risk damaging the head as it is tightened down, and has to push the liner down to the level common with the others.

The amount of material that can be machined off the cylinder head will depend on the condition of the head, and if it has been machined before to remove imperfections.

Quite large amounts of alloy can be machined from early model cylinder heads. After any such machining a cylinder head should always be 'try-fitted' to a block and valve-to-piston clearance checked at top dead centre (TDC). The cylinder head stud thread length should also be checked to ensure the head can be tightened down to the correct torque: the stud nuts are the domed type and the stud thread can bottom in the nut before it is fully tightened; alternatively ordinary through nuts can be used. The valve clearances are set by means of 'tappet shims' fitted to the top of the valve, underneath inverted cup tappets. Particularly when fitting improved camshafts, the tappet clearances should be measured and adjustments made as required.

A number of improved camshafts are available for these engines, including some manufactured by Autodelta, a company which originally made racing camshafts for the Alfa Romeo factory teams.

Original Alfa Romeo (as well as other manufacturers) big-end and main bearing sets are available for these engines in standard and a range of undersizes. If the crankshaft is reground to an undersize it should be rebalanced complete with the flywheel assembly and the connecting rods and pistons.

The oil pump fitted to the standard engine is perfectly adequate for racing use, as is the original sump which is well baffled so that oil surge problems are not encountered.

Both the carburettors and distributor originally fitted to the engine are perfectly suitable for amateur motor racing use. When fitted with improved camshafts the engine can utilize a larger size of carburettor.

A rebuilt Alfa Romeo unit should be carefully run-in before racing use is contemplated, and the workshop manual recommendations observed concerning the retorquing of the cylinder head nuts.

Ford 1600cc pushrod OHV engines

The 1600cc pushrod OHV Ford engine, and the units derived from it, is probably the production engine most widely used in amateur motor sport today.

Because the engine is used in a wide variety of applications there are a number of companies (besides Ford) who produce tuning parts for the unit. There is also a good secondhand market in Ford competition components.

The basic reconditioning and rebuilding of these engines is quite straightforward and clearly detailed in the appropriate workshop manual. The following notes will list and explain what tuning parts are available. The specific tuning that can be carried out on an engine will depend on the regulations for the class of race for which the vehicle is to be entered.

To help understand the modifications, the following should be noted. The Formula Ford engine is a unit derived from a 1600cc Ford Cortina GT pushrod OHV unit. The main differences between the standard unit and the Formula Ford regulation unit are: the crankshaft and reciprocating components are balanced, a small amount of cylinder head tuning is carried out, the water pump drive is converted to a small toothed belt drive, the lubrication system is converted to a dry sump system, and a racing exhaust system is fitted. Note also that in the following details of modifications which can be carried out to a Ford engine the term FULL-RACE engine refers to a 1600cc pushrod OHV Ford engine that is modified and tuned to the highest state possible.

Cylinder head. The standard Ford Cortina GT cylinder head is the basic motor sport head, which with some very minor modification is also the Formula Ford configuration head. Modified cylinder heads are available in higher states of tune, from simple gas flowing up to the fitment of large valves for full-race.

Valve springs. Springs are available in various loadings from the standard load of about 105lb up to full-race heavy-duty double springs of 210lb load. The cylinder head valve spring seats must be machined to prevent coil binding when fitting heavy-duty springs and improved camshafts. Note that the heavier the loading of the valve spring the greater the stress upon the valve train components; this creates higher wear and shortens component life.

Rocker gear. Strengthened valve rocker gear is available for use with heavy-duty springs or improved camshafts; a modified unit 'rolling rocker' gear is available for use with full-race specification heads, camshafts and springs. Lightened and balanced pushrods are available.

Cylinder block. The standard cylinder block can be rebored a number of times and most race class regulations allow a permitted maximum oversize dimension for this. When rebored to very near the maximum limit these blocks do have a likelihood of developing hairline cracks in the bores when used for motor racing. The block can be sleeved.

The 1600cc twin cam Ford block is often used for building up a very strong full-race pushrod engine. The twin-cam block has some additional strengthening webs and extra thickness in the casting when compared with the standard unit from which it is derived. The externally running camshaft belt drive and the camshaft is replaced with a normal camshaft and timing case cover: the engine is then built up as a normal pushrod unit.

Main bearing caps. Steel, as opposed to cast iron main bearing caps are available for these engines; they give increased support and rigidity to the crankshaft. To fit the caps the cylinder block must be line bored, that is, the block, with new caps installed, is very accurately bored through in line to the correct bearing surface size. Although perfectly adequate for most ama-teur motor racing use the original cast main bearing caps can fail at high rpm.

Crankshaft. The standard crankshaft, when balanced with all relevant assemblies, is perfectly satisfactory for low budget amateur racing use to moderate rpm: the standard crankshaft is also used in Formula Ford engines. When constructing a more highly tuned unit the standard crankshaft can be crack tested and tuftrided. Tuftriding (low temperature liquid nitriding) is a process developed in Germany about twenty years ago; it is a surface treatment which is applicable to all ferrous materials, the treated steels having an increased resistance to wear and fatique. A balanced, crack-tested, tuftrided crankshaft can be safely used at higher rpm than the standard component. The best full-race crankshaft for these engines is a machined steel component; such units will safely rev to over 10000 rpm. Steel crankshafts are very expensive.

Camshafts. There is a very wide selection of camshafts to suit all types of applications for competition work. The Formula Ford unit uses the standard Cortina GT camshaft. Where the regulations allow, a camshaft which improves on the GT unit is simple to install and costs little more than the original component. For full-race specification, camshafts, the models A6 or A8 would be used. These are high-lift camshafts and their fitting requires the use of strengthened valve gear and machining to the valve spring seats and valve pockets in the piston crowns. An engine fitted with such a camshaft has little power available below 6000 rpm.

Full-race, high lift, camshafts wear at a high rate because of the extreme profile of the cam lobes. To lessen this wear, full-race engines should never be run on tickover throttle or at low revs: on starting such engines should be run at between 2000 and 3000 rpm.

Cam followers. New cam followers should always be fitted in conjunction with a new camshaft.

Flywheel and clutch. The standard flywheel, balanced, is used on the Formula Ford engine and can be used on engines up to quite a high state of tune. The flywheel can be lightened by specialists and a lightened flywheel does increase acceleration.

Lightweight steel and alloy flywheels for use with sintered racing clutches are also available; they are essential for full-race engines.

The standard clutch is used in Formula Ford and many other forms of competition; it gives a balanced and progressive feel to the clutch action. Full-race clutches are made of copper which is sintered onto steel plates; they will transmit a great deal of power without overheating and burning. A sintered clutch is very direct in operation, being either in, or out, with hardly any feel in between.

Pistons. The Formula Ford engine is fitted with the original equipment standard pistons; these are suitable for most applications below full-race modifications. For engines tuned to full-race state, forged pistons, such as those manufactured by Cosworth, are available.

Connecting rods. The original equipment connecting rods, balanced, are used in Formula Ford and similar engines. The standard connecting rods can be improved for racing use by crack-testing, to eliminate any components with minute cracks which could develop and fail under high stress use, and then shot-peening; a process which stress relieves the metal. For full-race modifications, forged connecting rods are available from a number of companies.

Oil pump and sump. Various high pressure/high flow oil pumps are available for the Ford 1600cc unit, for use with the standard wet sump system. One of these 'improved' units should always be used for competition work.

The oil pump is located, with the oil filter, externally to the cylinder block; this makes a dry sump pump a simple exchange fitting. For dry sump operation the standard sump is replaced with a shallow dish sump which is scavenged by a front section of the dry sump pump, the supply and high pressure feed being operated by a separate rear section. For competition use wet sump operation, various baffled sump pans are available, many of increased oil capacity.

Carburation and fuel injection. There is a wide range of carburettors and inlet manifolds available. In amateur race class regulations where carburation is 'free', twin side-draught Weber or Delorto carburettors are the most widely used. On full-race specification engines the larger diameter throat Weber carburettors or fuel injection is used.

A few specialist companies can supply fuel injection equipment for the Ford engine; they are usually modified units manufactured for use on other engines. The Formula Ford engine regulations require the use of the original equipment Cortina GT single down-draught carburettor and manifold.

Exhaust systems. Exhaust systems tailored to suit any intended application are available. An improved free-flow exhaust system, together with improved carburation, will give a considerable increase in the power output of any engine.

Distributor and ignition system. Several companies offer competition-type distributor units for the 1600cc Ford engine and there is a wide range of electronic ignition units available. One of the most popular electronic units for competition use is the contactless Lucas Opus system.

6

The cooling and electrical systems

AN ENGINE, tuned or not, used for motor racing will attain substantially higher coolant temperatures than one used in normal road-going conditions. This is because the engine will be run at, or near, maximum rpm for between ten to fifteen laps of the circuit. Engines designed for sports cars will, usually, be less affected by constant high speed use than those designed for more sedate applications.

The temperature that the engine oil attains is as important as that of the coolant. At high temperatures the engine oil viscosity (thickness) reduces considerably. At low oil viscosity the engine oil pressure will drop and lubrication of the bearings will be impaired: the wear rate of moving parts is increased and there is a risk of bearing failure. For maximum engine performance and good life expectancy, the control of coolant and oil temperatures at near the optimum temperature is essential.

The radiator. The radiator should be in good condition, without any obstructions, and should be pressure tight. A radiator that has had considerable use will have deposits from the coolant partially restricting the tubes. To clean out the radiator, make wooden bungs and stop off the inlet and outlet tubes; fill the unit with a solution of soda or a proprietary radiator flushing agent. After soaking, reverse flush the unit with a hose pipe inserted in the lower outlet. It may take several soakings to fully clear an old radiator.

The radiator fins direct cooling air over the coolant tubes; large areas of flattened finning will considerably impair the radiator efficiency. The fins should be carefully cleared of debris between them with a stiff brush; an airline can also be used to blow out rubbish from deep in the radiator core. Areas of flattened finning can, with care, be straightened out with a soft wooden spatula or similar instrument.

The cleaned radiator should be pressure tested for leaks. A DIY pressure tester can be made using a non-pressure type flat radiator cap; these caps are fitted to the radiator on modern cars equipped with coolant recovery systems, the pressure cap being fitted to the expansion chamber. The flat cap should have a hole drilled in the centre to which is fitted a screw up type competition tyre valve.

To test the radiator, the outlet/inlet tubes are blocked off with short lengths of hose sealed with securely fixed bungs. The radiator is then filled with water and the tyre valve cap fitted; the radiator can now be pressurised through the valve with a foot pump: water will escape from any leaks until the pressure drops; the unit can be continually repressurised until all the leaks are identified. **Note:** the radiator should only be pressurised to a little over

Two simple DIY pressure testers

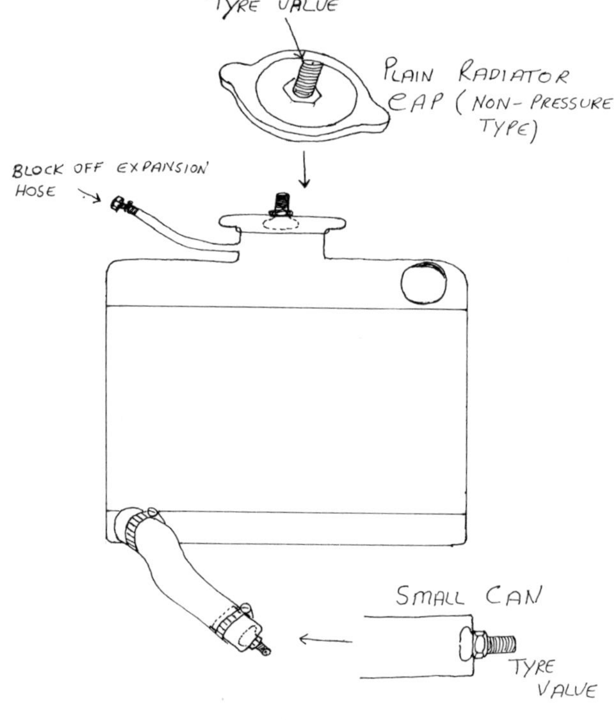

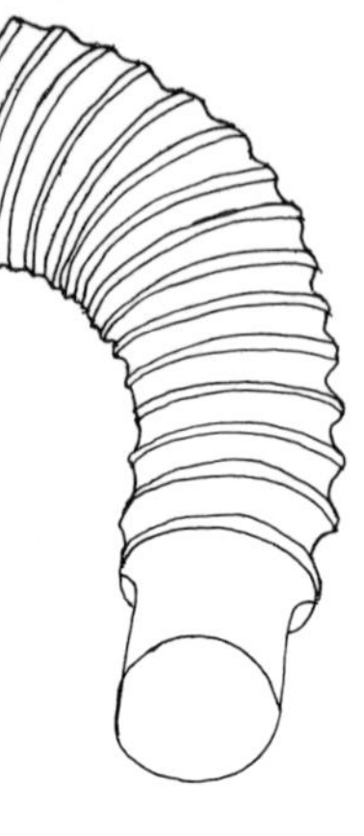

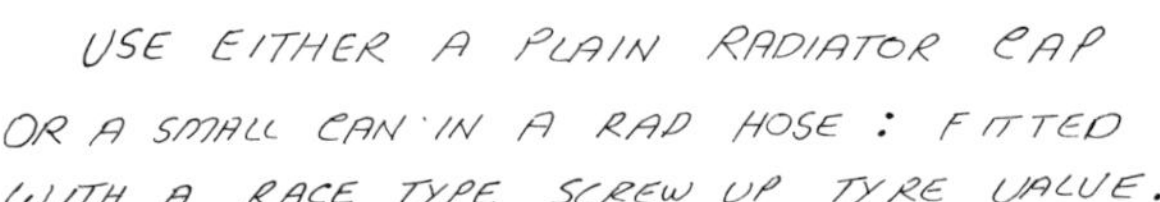

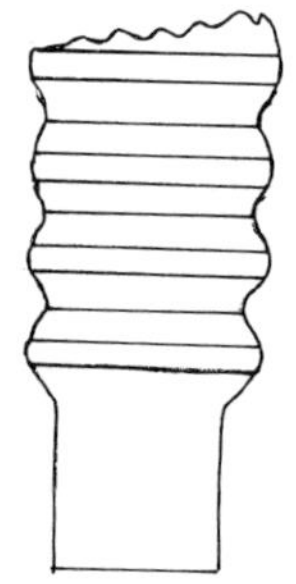

Convolute hose

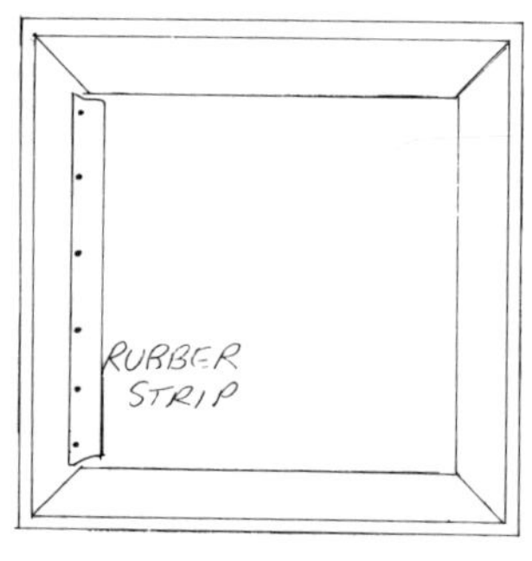

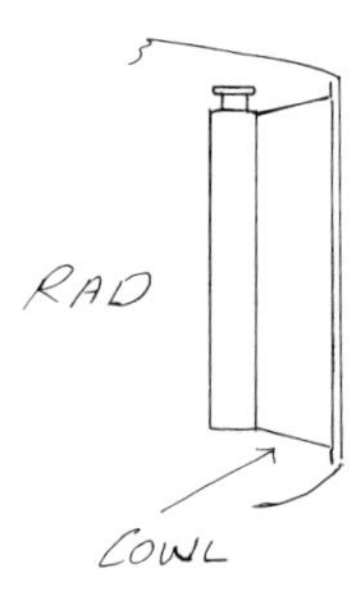

A radiator cowl

the pressure shown on the cap normally fitted to the the radiator.

During manufacture, radiators are soldered in a bath process. Small repairs to a damaged core can be made using a small gas torch with good quality solder and flux. When soldering, care must be taken not to allow excess heat to spread to the sound parts of the radiator core, and unsoldering further areas.

Radiator repairs can also be made with 'Araldite' and similar resin glues. The area to be repaired must be scrupulously clean when using such glues. So that the glue does not flow out of the repair area, becoming too thin over a hole, masking tape can be used to construct a small ridge around the site. Placing the radiator in a warm place, such as an airing cupboard, will decrease the setting time of the resin glue, and protect it from contamination by moisture. Always pressure test a repaired radiator.

The cleaned and refurbished radiator should be painted matt black; to aid heat dissipation.

An old radiator cap should be renewed, the flexible washer fitted to seal the base of the cap deforms with age and can cause the spring pressure release mechanism to jam. This, in effect, turns the engine into a pressure cooker, and as the pressure increases a radiator seam or hose will burst.

For racing, all the water hoses should be renewed. Due to the additional heat and stresses of competition, they should be replaced every season as a precautionary measure.

As a replacement for hoses that are difficult to obtain new, lengths of convoluted hose is available. It will bend to any shape required. Difficult hose shapes can also be made by cutting up new hoses. The pieces, at angles to suit specific requirements, being joined by short lengths of pipe and clips.

Many cars do not have a radiator cowling to direct airflow into the radiator. A cowl can be beneficial when racing in close bunched up traffic, in the hot exhaust air wake of other vehicles. A cowl can be made simply from thin alloy or steel sheet. The sheet seams may be pop riveted together and then sealed with race tape. Where the cowl meets the radiator, a lip seal strip of rubber, glued or riveted to it, will increase its effectiveness.

Water recovery systems. A closed-circuit water recovery system can be fitted to any engine. A recovery system is most useful in small emergencies, such as overheating the engine on the starting grid whilst being held up due to some starting delay.

The illustration shows the general arrangement of the recovery system components. A plain radiator cap is fitted to the radiator and the overflow pipe is connected to an expansion tank, which is fitted with a pressure cap. Heated coolant expands into the tank and syphons back into the radiator whenever the coolant temperature drops.

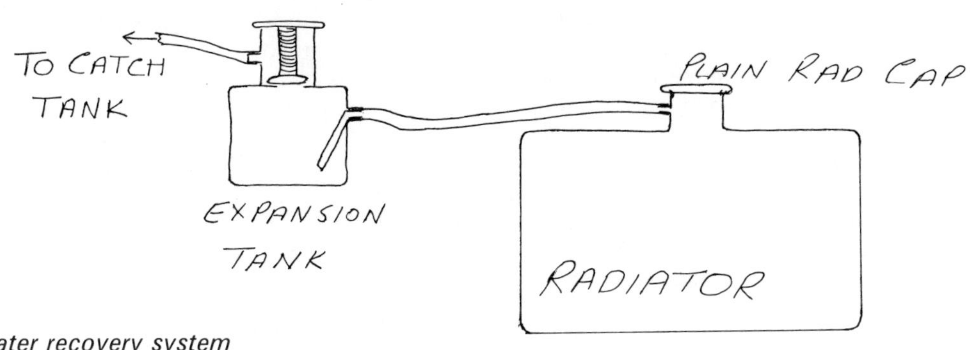

Water recovery system

A suitable expansion tank can be obtained cheaply from any scrapyard; one of the larger sizes should be used, such as those fitted to the Ford Transit van or similar vehicles. The expansion tank should be fitted as near to the radiator as possible. Reinforced nylon hose should be used for the connection between the expansion tank and the radiator; the thin rubber type of overflow hose will collapse as the radiator sucks water back on cooling. The expansion tank overflow pipe must be piped to the general catch tank.

The thermostat. During normal road use the thermostat's function is to restrict the engine coolant flow on starting the engine from cold, so that the coolant and engine attain the normal operating temperature quickly. The thermostat, even when fully open, still presents some restriction to the coolant flow around the engine. When using an engine for motor racing the thermostat is usually removed.

The radiator fan. Mechanically driven fans consume engine power and should usually be removed. Electric radiator fans, when operating, also consume power via the electrical load on the generator; however, under racing conditions sufficient airflow is generated to ensure that the fan is not switched on. If the vehicle is also to be used on the public roads a radiator fan will be essential; either an electric fan or a freewheeling mechanical type should be fitted.

The heater. The car heater unit can be usefully retained as an additional source of cooling when racing in very hot weather: switching the heater full on effectively places another small radiator in the coolant circuit. The windscreen demisting function of the heater can also be very useful when racing in typically British weather. Some racing driver/mechanics prefer to remove the heater unit because: it is something else to go wrong and the hose and connections could be a source of coolant leaks; it is unnecessary weight; it interrupts the smooth coolant flow through the engine and radiator; and it can cause air locks.

Overheating problems. Under racing conditions an engine, even though only mildly tuned, may develop coolant overheating problems. Provided there is sufficient coolant capacity there are two areas which could cause the problem:

1. The water pump may be turning too fast at maximum rpm. If the coolant flow is too high, heat transfer from the metal to the coolant is impaired. A larger size pulley should be fitted to the water pump: this will reduce the pump rpm.

2. Air locks may be forming in some parts of the engine water galleries. At high temperature some engines can form 'steam pockets' in awkward corners where the water flow is restricted. To cure this problem a 'swirl pot' should be fitted. In modified engine installations, where considerable water pipe 'plumbing' is required, a swirl pot is essential to de-aerate the coolant.

Oil coolers. In amateur motor racing, with races of short duration, engine oil temperature problems are not often encountered. Overcooling of the engine oil (and water) is as detrimental to engine performance as is overheating. However, depending on the model of car and the state of tune of the engine, fitting an oil cooler may sometimes be beneficial. Observation of the engine oil temperature and pressure under racing conditions will determine if an oil cooler is required.

Oil cooler installation kits are available for a wide range of engines. For engines fitted with a spin-off oil filter, a cooler assembly is available which consists of an alloy plate, with inlet and outlet for oil flow, which is fitted between the oil filter and the engine mounting. For engines with non-spin off replaceable oil filter elements a spin-off filter converter will often allow the use of the former cooler installation as well. Engines for which converters are not available can sometimes be fitted with an oil cooler if the oil feed pipe to the filter is

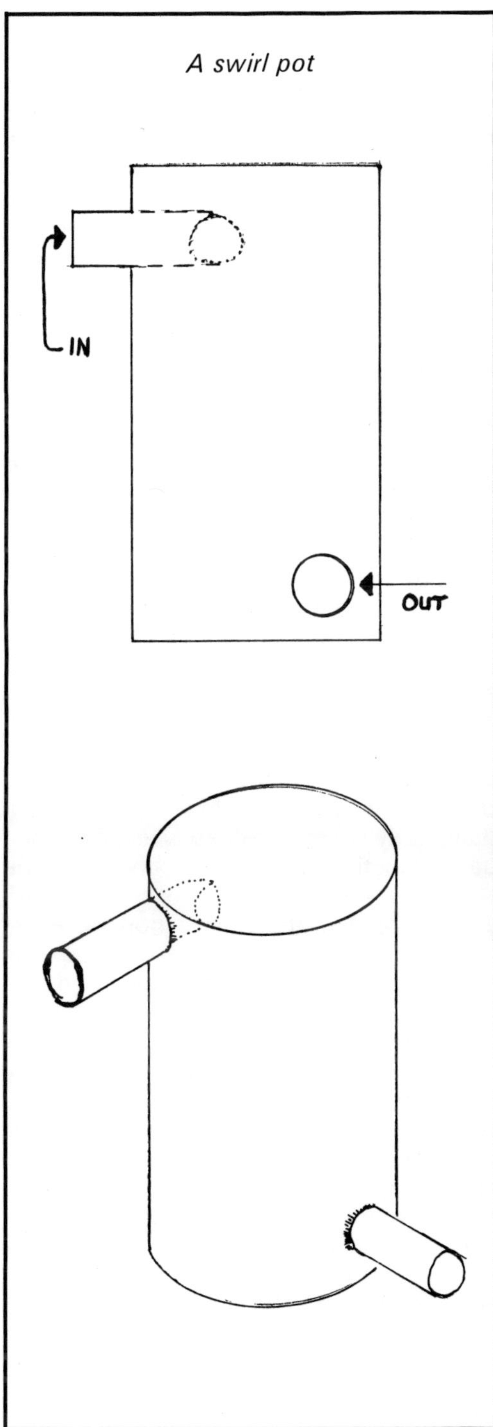

mounted externally: the filter pipe is cut and extended to provide the flow and return to the cooler.

The oil cooler should be mounted in a position of clear air flow and, if mounted in tandem with the water radiator, then in front of it, not behind in the warm air flow. The oil cooler flow and return hoses should be high pressure oil hose, and preferably of the metal braid protected type. Apart from racing car parts specialists, high pressure oil hose can also be obtained from suppliers of hydraulic equipment and fittings.

NOTE: because of the removal of the coolant thermostat and other modifications, the racing engine should always be started and thoroughly warmed up to near optimum temperature before the car is driven on to the circuit or the engine stressed in any way.

Electrical systems

The wiring diagram for your particular model of car will be found in the relevant workshop manual. The race class regulations will state what electrical equipment, such as lights, are required. If the car is to be used only for motor racing any extraneous wiring and electrical items can be stripped out.

Most problems with vehicle electrical systems arise because of the very low voltage involved. Corroded connections, poor earth returns and earth straps to the engine unit, together with rust, often combine to inhibit the passage of current. All connections and earth points must be scrupulously clean and maintained so: a smear of vaseline over earth points and battery studs will usually keep them bright.

In racing cars, electrical connection problems are often caused by vibration. Whenever possible, push-on connectors should be converted to a captive ring type, secured by a nut over a stud or bolt. Crimp connections can be made positively secure by additionally soldering the wire to the crimp, and the push-on connector, with one spot of solder, to the spade.

Cable/wire. Low voltage electrical systems run at comparatively high current values. When installing new wiring, or rewiring existing components, it is essential to use automotive grade wiring cable of the correct section for the loading required. Household flexible wiring cable, designed for use with a 75 watt bulb at 250 volts, will melt if used to connect a 75 watt bulb at 12 volts in a motor vehicle: the cable would overheat at the considerably higher current load.

As an added precaution against wiring failure, some racing car constructors always run twin wires to all essential components, such as the ignition system.

Rear facing red warning light

Apart from individual race class regulation requirements (such as that Classic Saloons must have all the normal car lights) the only mandatory electrical requirements of the General Racing Regulations are that the vehicle be fitted with a rear facing red warning light and emergency on/of switch, and that the battery be fully enclosed.

The red rear warning light is used when racing in conditions of rain or poor visibility. The correct operation of the light is always checked by the scrutineers at every pre-race vehicle check.

A red rear fog light, or rear hazard add-on brake light type fitting, is the most usual type fitted. The light should be installed on the centreline of the vehicle body. On saloon cars the usual fitting position is on the rear panel above the bumper line, or on the rear parcel shelf provided the light is clearly visible through the rear window.

The light must be wired to a clearly marked and accessible on/off switch, which the driver can operate whilst strapped in his seat.

The full wiring system. In race classes where the normal electrical system and components of the car are retained, all connectors and components should be checked regularly for soundness and signs of failure due to vibration. In vehicles where the maximum engine rpm available has been increased (above its original design) the dynamo, or alternator, pulley size should be increased to reduce the components' speed at maximum engine rpm. At high rpm, dynamo or alternator armatures can disintegrate.

Any item of electrical equipment in use places a load upon the engine, using up horsepower. An alternator only causes an engine load when it is charging. This effect can be eliminated by wiring an on/off switch in the field pole line of the alternator (the terminal marked F). The alternator can then be switched out of circuit whilst racing.

Total-loss wiring systems. In race classes where no electrical equipment is required (other than the essential ignition, starting and warning lights) a total-loss wiring system can be employed. Apart from ignition, starting and warning lights, all other wiring and components can be stripped off the vehicle. The car is then run on the battery charge alone. It is essential that a fully charged battery is installed for a day's practice and racing, and that the engine starter is not used excessively. All amateur and professional, short distance, single-seater racing cars are wired this way, as are many of the fully modified cars in other classes.

Auxiliary power socket. Whilst not essential, when racing without a battery charging system, facility for an auxiliary plug-in battery boost is useful. Jump leads and crocodile clips are too slow in use and are dangerous: a short circuit could give a spark that might set the whole car alight.

An auxiliary plug and socket must be non-reversible, so that polarity is not accidentally switched. There are several plug and socket sets on the market made specifically for racing car use; most have the facility to pull out should the driver accidentally move off having forgotten to

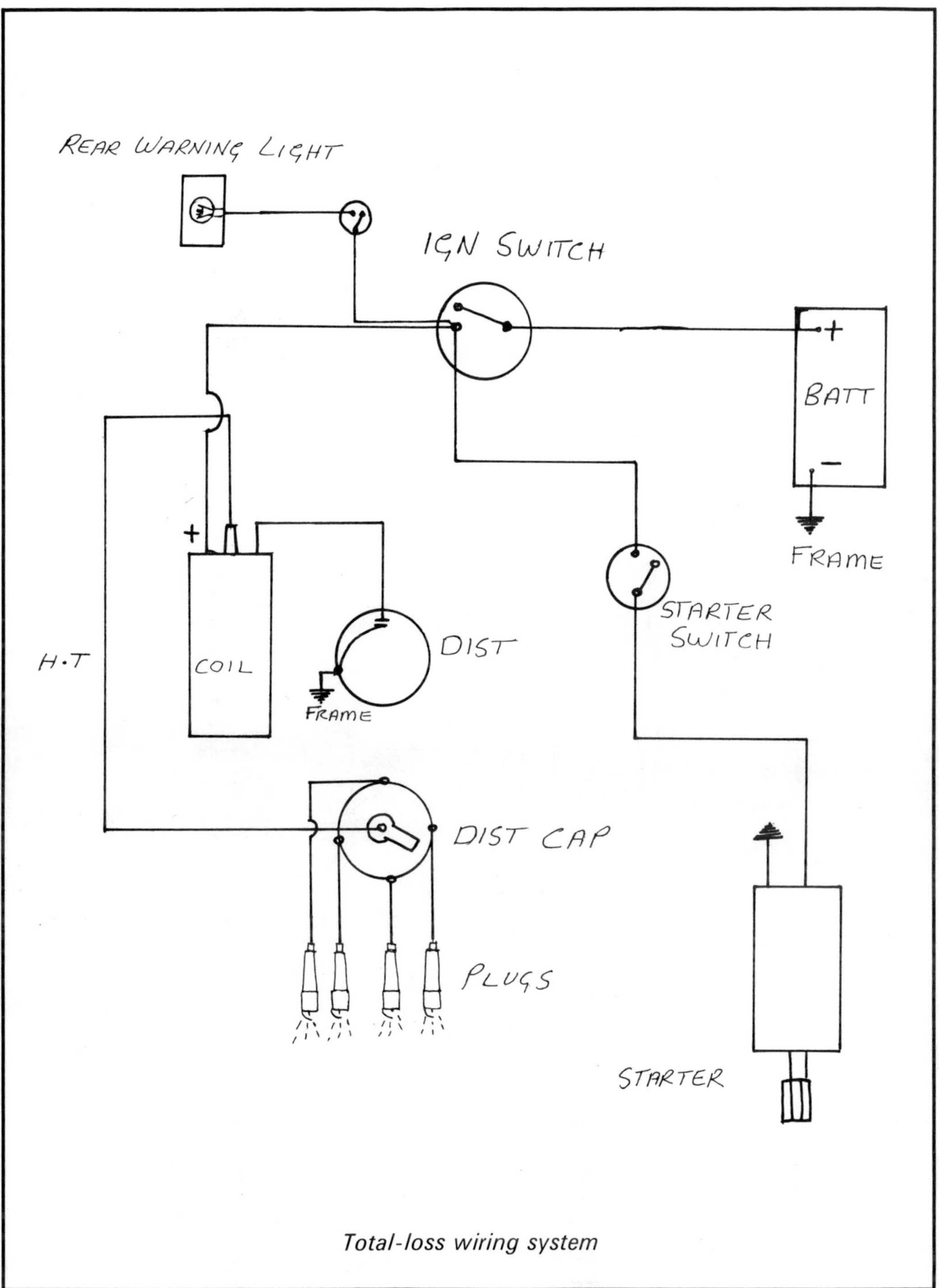

Total-loss wiring system

unplug the battery.

It is very important to use really good quality large diameter flexible cable for the auxiliary socket 'flying leads'. If this is not used, when the cable gets hot, it will not conduct the current sufficiently well and one could mistakenly think that the battery is discharged.

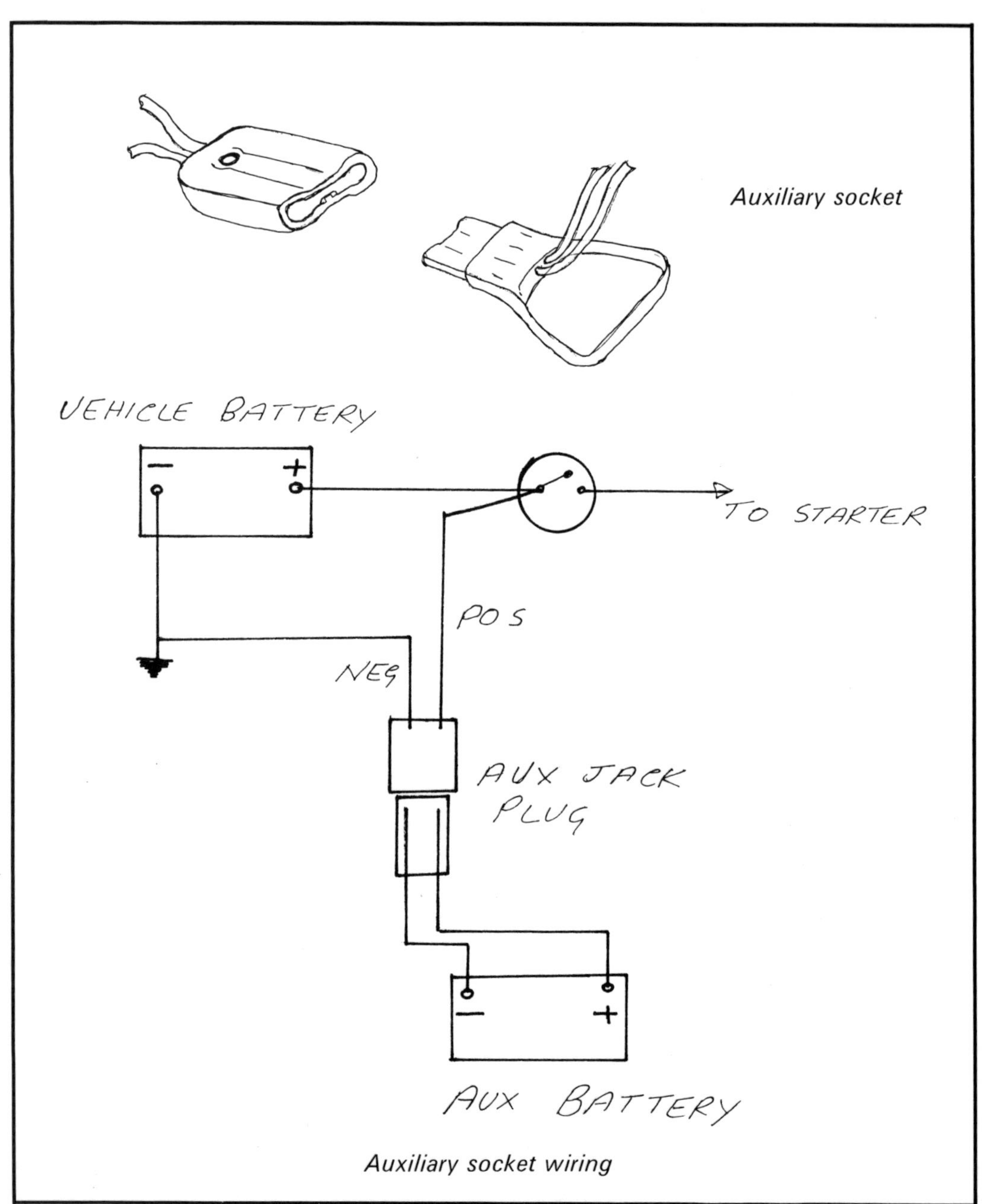

Auxiliary socket wiring

7

Single-seater and sports racing cars

SINGLE-SEATER and sports racing cars almost always cost more than £1000. There are, however, two ways in which you can purchase such a car on a low budget: by forming a racing partnership with a friend, and sharing the purchase price; or by buying a car in poor condition in need of total renovation. Single-seaters are open cockpit vehicles with minimal bodywork and no wings or mudguards covering the wheels; there is only one seat, usually positioned on the centre line of the car.

Sports racing cars are sports cars constructed specifically for motor racing; there must be space for the fitting of two seats, although generally only one is fitted (to save weight) and the wheels are covered with wings or mudguards. The driver's seating position is usually to the right-hand side of the centre line of the vehicle. Being constructed solely for motor racing use neither type of vehicle is usable on the public highway and the chassis to ground clearance is often less than two inches.

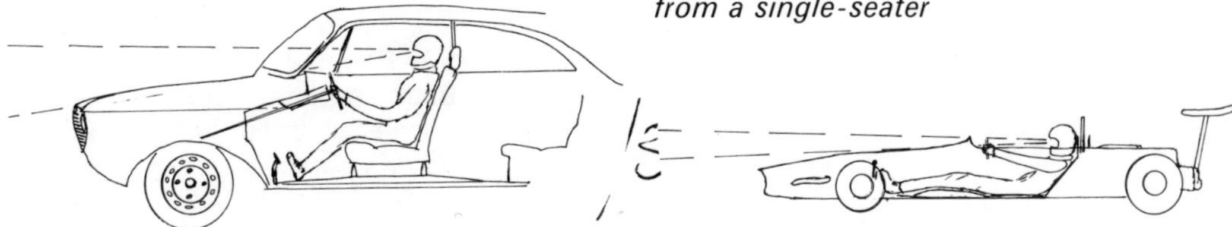

There is quite a different view of the track from a single-seater

Construction

In order to obtain a centre of gravity as near the road surface as possible, single-seaters and sports racing cars are constructed to a low profile with ground clearance at the minimum allowed by the formula regulations. Most of the suspension components are constructed so that they are adjustable; this allows the steering and handling characteristics of the car to be 'set-up' to suit the driver and the circuit conditions. The weight of the vehicle is reduced to a minimum, taking into account rigidity and safety, and this increases the vehicle speed attainable for an engine of given power output. These vehicles use various forms of racing tyres which give an increased grip of the road surface at the expense of tyre life.

The body-chassis structure can be a space-frame, a monocoque, or a combination of both with a one piece 'tub' (main body, centre section) and sub-frame extensions for the suspension and engine components.

There are racing formulas for front and rear engined cars. Most formulas specify a particular engine model and the configuation to which it must adhere; for example. the Formula Ford engine, which as well as Formula Ford cars is used in Clubmans B and Monoposto Kent formulas.

The Formula Ford engine is a pushrod 1600cc Ford Cortina GT engine, with fully balanced components, a small amount of cylinder head tuning, a more efficient exhaust system and a toothed belt water pump drive together with a dry sump oil supply conversion. The engine is effectively very little different from the normal road-going unit it is derived from.

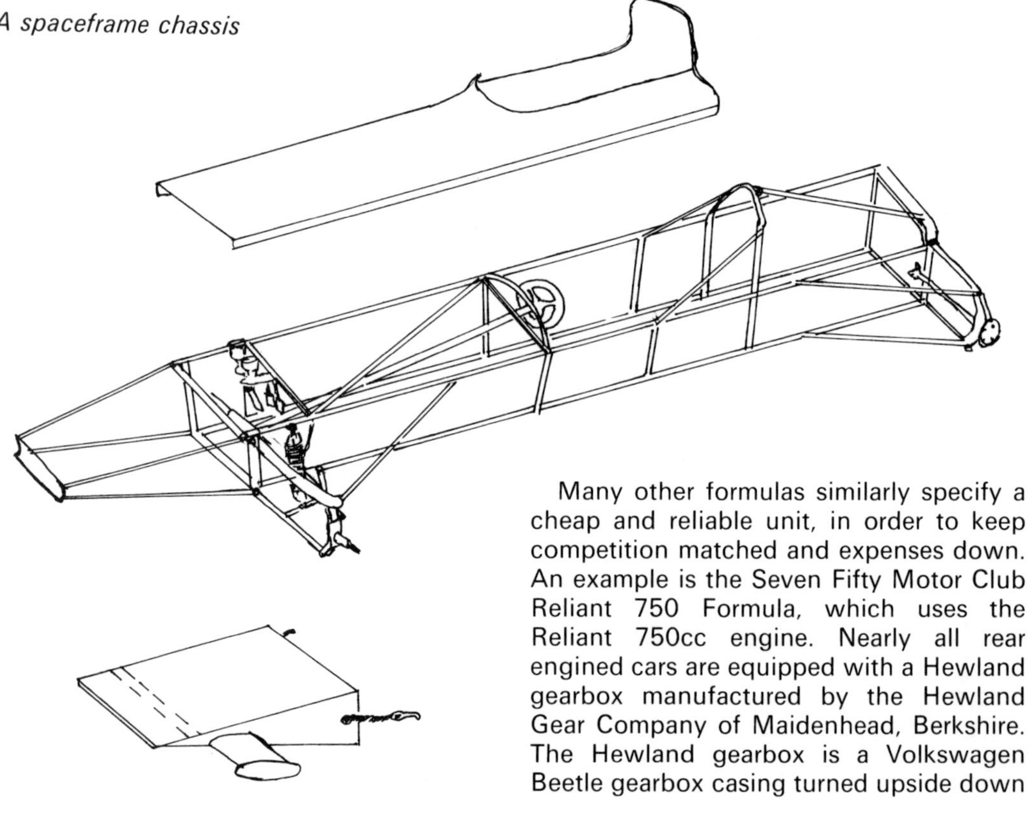

A spaceframe chassis

Many other formulas similarly specify a cheap and reliable unit, in order to keep competition matched and expenses down. An example is the Seven Fifty Motor Club Reliant 750 Formula, which uses the Reliant 750cc engine. Nearly all rear engined cars are equipped with a Hewland gearbox manufactured by the Hewland Gear Company of Maidenhead, Berkshire. The Hewland gearbox is a Volkswagen Beetle gearbox casing turned upside down

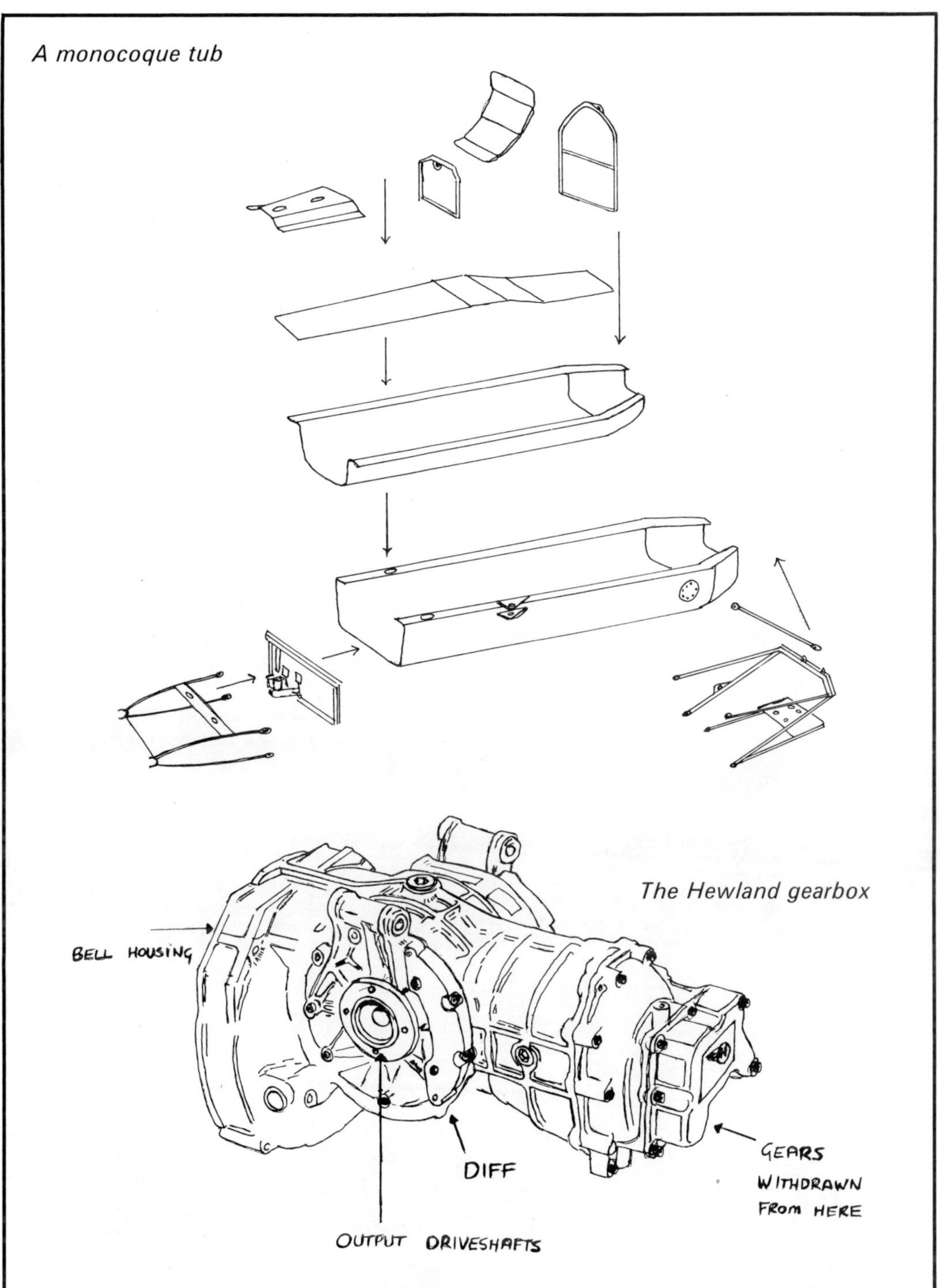

A monocoque tub
The Hewland gearbox
BELL HOUSING
DIFF
GEARS WITHDRAWN FROM HERE
OUTPUT DRIVESHAFTS

and fitted with straight-cut gear sets. The gearbox rear end case can be removed very quickly and the gear ratios changed with ease: the engine's maximum performance can therefore be 'geared' for any particular circuit.

Front engined cars use a variety of normal road car gearboxes coupled to the engine in the normal manner; consequently the drive train arrangement for a front engined car is substantially cheaper to buy than the specialised Hewland gearbox.

Clubs

The next section of this chapter gives details of motor racing clubs for single-seaters and sports racing cars, with some individual examples of some low budget single-seater and sports racing cars in more detail.

All the motor racing clubs which organize races for this type of car cater for amateur racing drivers, often with a low budget: most drivers do all their own preparation and maintenance work on their cars.

Formula: Reliant 750 Formula

Club: The Seven Fifty Motor Club.
Contact. The general secretary, Mr Dave Bradley, 16, Woodstock Road, Whitney, Oxon. OX8 6DT. Telephone (office hours). Whitney 2285. The Seven Fifty Motor Club also organises races for: Formula 4, Formula V, Formula Ford, all single-seaters; Sports racing cars and sports cars. *Example vehicle: Reliant 750 Formula car.* Most of these vehicles are manufactured by amateur constructors. The major requirement of the formula for the chassis being two main longitudinals of 50 x 50 x 1.5mm, square steel tube of a minimum continuous length of 52in.

The engine must be a standard Reliant overhead valve unit of up to 748cc. There is an additional bore wear allowance dimension. The standard Reliant crankshaft must be used as must the standard cylinder head and ports. A race profile camshaft is allowed.

The gearbox must have a maximum of four forward gears; aluminium alloy gearbox bellhousings are forbidden. A proprietary live rear axle must be used and limited slip differentials are not allowed.

Reliant 750 Formula (courtesy 750 motor club)

Formula: Formula V

Club: The Seven Fifty Motor Club.
Example vehicle: Volkswagon Formula V.
These cars are based on the 1300cc
air-cooled Volkswagen engine and gearbox. The regulations specify the use of a
considerable number of suspension and
steering parts from the Volkswagen Beetle
car. The degree of engine tuning is very
limited and most components must be in
standard condition.

There are a variety of American and
European-manufactured chassis available
and a kit consisting of just the basic frame
can also be purchased.

Formula V (courtesy 750 motor club)

Formula: Monoposto 'Kent'

Club: The Monoposto Racing Club.
Contact. The honorary and competition
secretary, Ms Suzy Livingstone, 2 Juniper
Road, Langley Green, Crawley, Sussex.
The Monoposto Racing Club also or-
ganizes races for single-seater cars with
1600cc, 2-litre and 1-litre engines, historic
Formula Junior and Formula 3 cars.
Example vehicle: Monoposto Kent-Merlyn.
The car pictured belongs to Chris Winter of
Southampton and was formerly a Merlyn
Formula 3 car. Manufactured in 1973, it
was originally fitted with a Ford BDA
twin-cam engine.

The Monoposto 'Kent' formula allows
the use of any single-seater racing car
chassis manufactured over five years prior
to the start of each racing season, or a
home-built chassis of any age. The
Formula Ford 1600cc engine is stipulated.

Chris Winter bought the Merlyn 'rolling
chassis' (a chassis and suspension with
wheels less engine and gearbox) for £600

in 1982 and built the complete car for £1200. Such chassis at similar prices are always available.

Chris Winter in his Monoposto car

Formula: Clubmans 'B' Sports Racing Cars.

Club: The Clubmans Register.
Contact. The chairman Mr Chris Hart, 23 Ranmoor Court, Graham Road, Sheffield, SL10 3DW.
Example vehicle: Mallock.
The formula caters for those who wish to build their own chassis, and forty-five different makes of clubmans' chassis, in varying numbers, have been constructed since the club's inception. The most successful have been those constructed by Arthur Mallock, who has manufactured chassis in large numbers.

There are two race classes, A and B, with the B formula specifying the Formula Ford 1600cc engine to keep construction and maintenance costs down. The A class permits more highly tuned engines of up to 1700cc capacity.

Formula. Pre-74 Formula Ford.

Club: Several of the national motor racing clubs organize races for Pre-1974 Formula Ford cars. On joining one club, such as the BRSCC, you will be eligible to enter races organized in the race series by the other clubs.
Vehicle Details: Formula Ford chassis have been produced in quite large numbers since the inception of the formula. With technical advances in suspension design occurring almost every year, early model chassis are available quite cheaply.

All Formula Ford chassis are of space-frame construction, the early examples in particular being very simple to dismantle and rebuild.

The formula engine is the standard Formula Ford unit, derived from the 1600cc pushrod OHV Ford Cortina GT

Two examples of Clubmans Cars

engine. A straightforward unit to rebuild and maintain it is very reliable if used within its limits. Virtually all early Formula Ford cars are equipped with the Hewland four-speed gearbox, although some can be found with a simple inverted standard Volkswagen gearbox.

A considerable business has developed dealing in secondhand Ford car parts: the availability of cheap used components makes rebuilding even a severely damaged car a reasonable proposition.

Learning the art of race driving

THERE ARE FOUR ways in which one can gain some experience of driving a racing car on a race circuit before entering a race.

1. Racing schools. Many race circuits throughout the country have a racing drivers' school based at them. The schools usually offer a one-day introduction course, costing about £50.00, and a full course of three or five days, costing about £300.00.

The one-day introduction course consists of several laps of a circuit in a saloon car, first with an experienced racing driver at the wheel, showing the novice the basic cornering technique, followed by the novice taking the wheel. The instructor makes a few notes of the novice's abilities.

A short classroom session follows, in which the novice driver's performance on the circuit is discussed and further techniques explained. The final session consists of further laps of the circuit which are driven using a Formula Ford single-seater racing car. An introduction course such as this will teach the rudiments of race driving and introduce the novice to a circuit under safe controlled conditions.

2. Motor club test days. A number of motor clubs and motor racing clubs organize racing circuit test days, during which a racing circuit is booked for the sole use of their members. The aim of the test days is to allow club members to sample

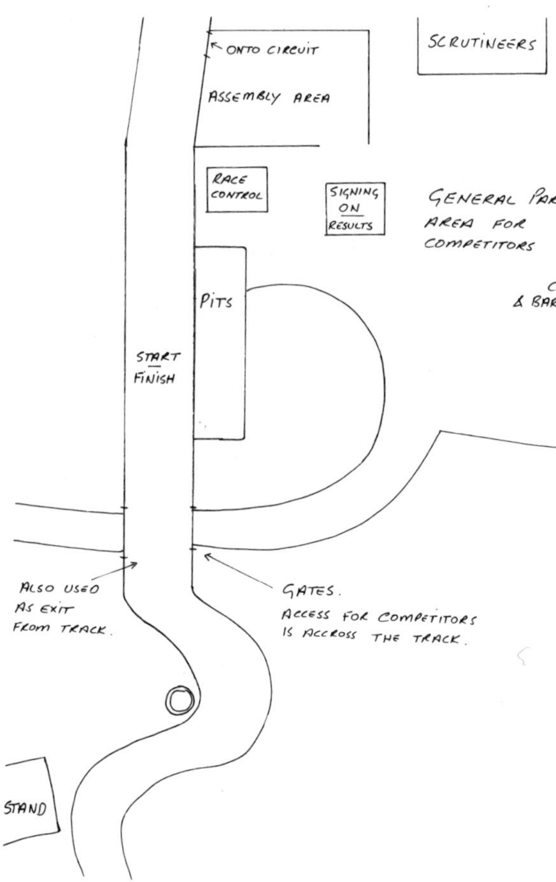

The paddock layout at Thruxton circuit

driving on a race circuit under safe and controlled conditions, but whilst actual racing is not permitted.

Club members use their normal road cars on the circuit and whilst approved crash helmets are required, many of the other general requirements, such as roll bars, are waived. Test days are a useful method of experiencing driving on a circuit, using all the road available, and with no traffic coming in the opposite direction.

3. Race circuit general practice days. All racing circuits have a number of general practice days each year, on which any racing driver may take his car and practice driving on the circuit. These general practice days are also called 'test days' because drivers and mechanics also use them to test a racing car's performance and make and test adjustments to the vehicle.

The RAC General Racing regulations apply to test and practice days: the vehicle must be examined and passed by a scrutineer; roll bars must be fitted and all safety requirements for racing cars must be complied with; the driver must hold a racing drivers licence and current medical certificate which must be presented to the organizers; and helmets and safety clothing must be worn.

The charge for practice days varies at different circuits, usually being between £30.00 and £45.00.

An important point for the novice amateur racing driver to consider about practice days is that many different classes of car will be present on the circuit. Inexperienced drivers driving on the circuit in company with cars capable of perhaps twice their maximum speed, can find themselves in dangerous situations. Personally I would argue that it would be safer for a novice driver to enter a race for his own vehicle class run by his own club. In such a race his fellow competitors would be well aware of his novice standing, and the number of vehicles on the circuit would be far less than can be encountered on a circuit practice day.

When a driver has gained some experience, circuit practice days are an invaluable aid to exploring and refining his car's performance, away from the heat of competition.

4. Hill climbs. Since only one car at a time attempts the course, hill climbs can be a safe and useful method of trying out a car and gaining experience. Although serious 'hill climbers' are very competitive drivers, hill climb meetings are very friendly gatherings of enthusiasts and the novice driver can learn much, pitting himself and his car against the course and the clock without the distraction of other drivers on the circuit.

Entering your first race

The racing calendar, the dates for the season's races, is decided by the organizing clubs and race circuit owners each winter; it is published each spring. Notification of individual race meetings and entry forms are posted by each racing club about one month before the meeting is due to take place.

You should complete and return the entry form for the race you wish to enter as soon as possible after receipt: at many circuits the number of cars allowed to compete in each race is limited, typically to between twenty and thirty entrants. The return of your accepted entry will contain two car passes for the paddock and five personnel entry tickets. The car passes are for your tow car or mechanic's car, a racing car towed or carried on a trailer to the circuit does not require a pass.

It is advisable to arrive at the circuit early. The driver's first task is to sign on, signing a form accepting the conditions of the race organizers, and present the racing drivers licence and medical form for examination; you will be given a race timetable programme and a ticket to hand to the scrutineer. Although the race acceptance form will show a time for scrutineering for your race number in the programme, entrants can often be examined early and this will leave yo with free time to familiarize yourself with the circuit layout: Affix the scrutineer's pass ticket to the racing car.

If you have not driven on this particular circuit at a club or test day you should go to the office of the clerk of the course, where a race official will tell you about the layout of the race circuit and of any important points to watch out for. Some circuits also arrange for new drivers to be driven around the track in a saloon car before the day's practice starts.

In ample time, before the announced practice time for the race, the car should be started and the engine run to warm up the oil and water to normal working temperature. The morning practice session will decide the grid positions of the competing cars: all cars must complete three laps of the circuit for practice in order to qualify for inclusion in the race.

The public address system will call competitors by practice session number to the assembly area, the pits, or a separate waiting area; you should ensure that you are changed and wearing your racing overalls well before the time shown on the instruction sheet: race days can be chaotic and times altered suddenly.

As a novice in your first practice and race you should concentrate on learning the way around the circuit at speed, whilst ensuring you do not get in the way of the experienced drivers; they will be competing for a good position on the grid. The lap times attained by experienced drivers in the practice are often higher than those which will be attained during the actual race.

After the practice the car should be examined, any adjustments that are necessary should be made, and the oil and water topped up if required. Leaving these tasks until just before the race often leads to mistakes. About half an hour after the practice session the results will be published and posted up at the Race Control Office; these will show each competitor's best lap time and the grid positions for the race.

A drivers' meeting will often be called at the race control office before the start of racing in the afternoon. At the meeting the Clerk of the Course will give drivers details of any procedure to be adopted for the races, such as the number of warm-up laps and whether the method of starting the race will be by flag or lights.

The car should be started and the engine warmed up well before the stated start time of the race: since amateur race meetings often have a large number of races at each meeting they are run to a tight time schedule. On parking the car in the assembly area you should be ready to put on your helmet in good time before the marshall starts waving the cars for your race onto the track.

On the track, cars first make a tour of the circuit coming round to line up on the start position grid. Grid marshalls wave a yellow flag to slow drivers and direct them to their exact position on the grid; engines are left running. On the grid, mechanics are allowed to attend the cars for a few moments before the 'two minute' board is held up by the Chief Grid Marshall at the front of the grid. The 'one minute' board is shown next and after mentally counting thirty seconds drivers put their cars into first gear and rev the engines ready to start. The chequered flag is dropped (or the red lights change to green) and all cars start off. If a warming-up lap has been scheduled the cars drive around the circuit for one lap in grid order and reform on the grid, as soon as the grid is reformed the flag or lights will again be given and the race starts. Should you stall the car on the grid the correct procedure is to put your right arm out of the window and hold it pointing straight upwards; grid marshalls will come to your assistance.

As a novice driver in your first race you should be particularly careful about observing other cars in your rear view mirrors. Even in a ten lap race the leaders may well go fast enough to come round and overtake you after five laps or so. The

Your mechanic can signal any little problems you might not notice

important aim in your first race is to drive the full distance and learn from the experience.

Setting up the car

The term 'setting up the car' refers to making such adjustments as are possible to the car in order to make the best of the vehicle's performance at each particular racing circuit. There are three aspects of setting up the car which when combined will enable you to obtain the best performance from the vehicle at each track.

1. The first aspect is the identification of faults with the vehicle under racing conditions. Fuel surge whilst cornering at speed may become apparent: only sufficient petrol to last the race distance, plus a small margin, is usually carried; however, you may find that twice this amount is necessary in order to keep the fuel pickup tube covered whilst cornering. Similar problems, such as the need for change in the engine sump oil capacity, may only show up under racing conditions at some circuits.

Always signal which side of you an overtaking vehicle should pass

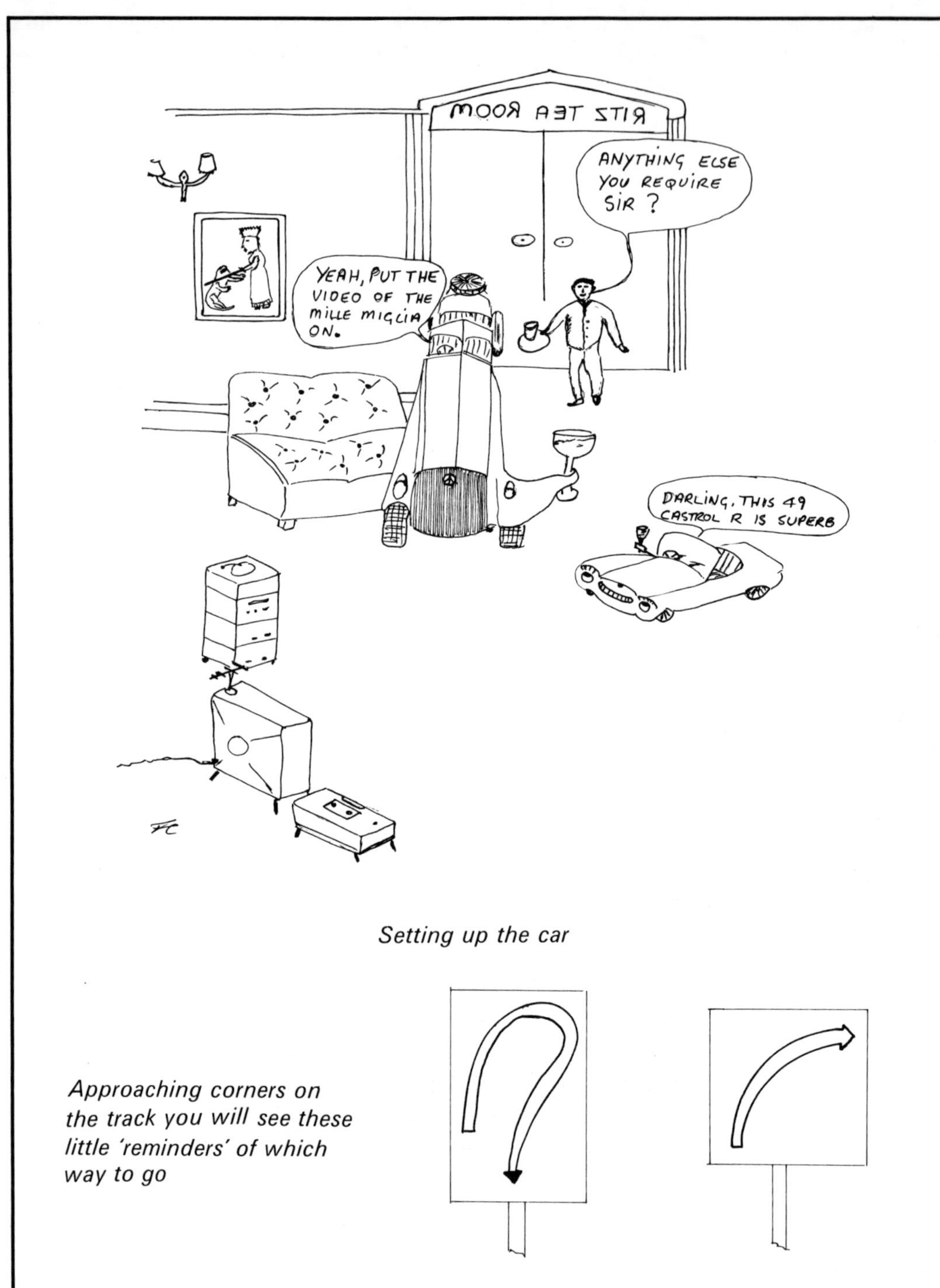

Setting up the car

Approaching corners on the track you will see these little 'reminders' of which way to go

During your first few races you should make a point of noting any faults in the car's performance or handling, so that you can investigate adjustments or modifications to mitigate the problem. Most motor racing clubs have a club photographer who, apart from normal souvenir photographs, will also take pictures of vehicles whilst they are negotiating particularly difficult corners or sections of the track. These photographs can often be used to identify particular chassis handling problems: it is very difficult for an inexperienced racing driver to determine the actual behaviour of a car's suspension whilst cornering at speed.

2. The second aspect of setting up the car is to make such adjustments as are possible to the vehicle to suit the conditions prevailing at a circuit for each particular race.

Adjustments of this nature would include: reducing tyre pressures when racing in the rain; adjusting shock-absorber damping to suit bumpy or smooth track surfaces; increasing tyre pressures at front or rear only, to affect oversteer. Any small adjustments which can be made to tune a car for a particular circuit will increase the vehicle's performance and make it easier to drive.

3. The third aspect of setting up the car is concerned with the driver, rather than the vehicle. Part of the process of gaining maximum performance from the car is that the driver drives the vehicle – sympathetically.

If the vehicle, for instance, oversteers badly on a particular corner you could change your method of driving through that corner. As a general rule the best lap times result from a smooth driving style.

At each race you should try to gain some 'feel' for how the car's suspension and chassis performs at that circuit, and after adjusting the car – adjust your driving technique to make the best of the car.

9

Basic race driving techniques

THE PURPOSE of this chapter is to outline the basic driving techniques used in motor racing, and to illustrate how they differ from the techniques of fast driving on a public road.

With an understanding of the fundamental principles, a novice racing driver should be able to venture on to a circuit for the first time in practice and tour the track at speed and in safety.

To achieve a fast lap time the full width of the track surface must be used when cornering, and it is in such aspects of racing driving, with no traffic coming in the opposite direction, that driving techniques differ from driving at speed on the ordinary road. Concentration in racing driving is very important.

The racing driver must develop the ability to concentrate solely on his driving, whilst ignoring extraneous information to the matter in hand. An aid to concentration is to develop the habit of looking at the vehicle gauges (oil pressure-water temperature) each time at one particular spot on the circuit; for instance, whilst covering a straight section. This will help the driver develop a rhythm in his driving at each particular circuit, rather than losing concentration to look at the gauges at different odd times.

The vehicle's gauges will be easier to examine quickly at speed if they are turned in their housings so that the indicator needles are pointing straight up, in the driver's line of sight, when at the correct setting. The engine revolution counter in particular should be positioned so that the needle is straight up (vertically) at maximum engine rpm. A gauge pointer that has dropped back from the vertical position is easier to spot with a quick glance than looking to see if it has reached a correct setting.

Racing drivers use the rev counter to assess vehicle speed and performance, because speedometers are too inaccurate, and they do not react fast enough to accommodate the speed changes encountered on a racing circuit. An important factor in motor racing is to keep the engine operating within the best power band of rpm. Whilst cornering, it is important to select the gear that will allow the vehicle to drive around the bend at near maximum rpm, so that on changing up the engine rpm do not drop out of the power band. Accordingly, the rev counter is used to judge which gear to select for any given section of a circuit.

Steering

The basic steering wheel hand position is shown in the illustration. Moving the steering wheel through the hands whilst driving at high speed can lead to disorien-

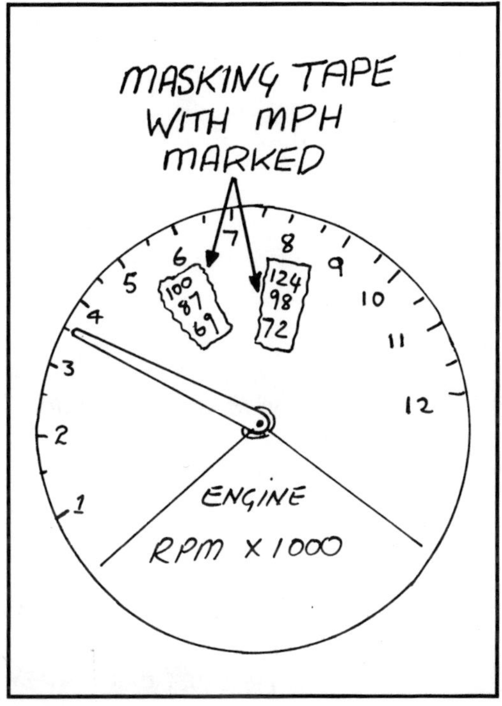

Until you get used to a rev counter stick speed reminders on the dial

Hand position on the steering wheel

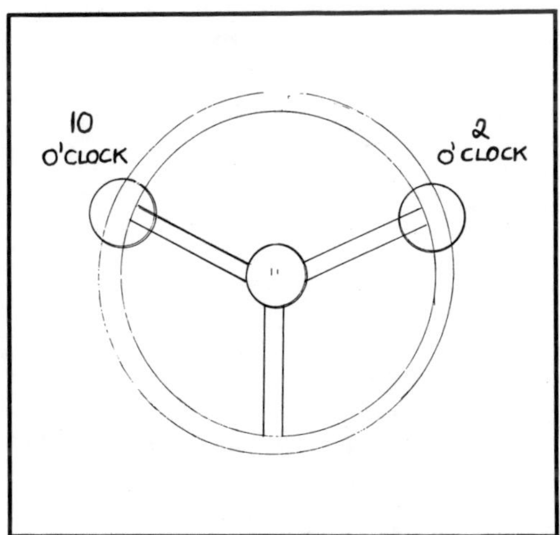

tation of the straight ahead position (in relation to steering wheel position).

Provided the vehicle controls can be so adjusted that the driver's arms are fairly straight relative to the steering wheel, most bends on a circuit can be driven round without moving the hands on the wheel rim. In racing driving, it is preferable to cross the arms turning the steering wheel on a tight corner, so that the hands need only be turned back to the starting position to ensure that the car is again steering straight ahead.

It is important not to let the castor action of the vehicle (self straightening action of the front wheels to the steering wheel) take over the steering on exiting a bend: at speed, over-correction is likely to occur and the driver will have to make further steering corrections.

If your vehicle exhibits a strong castor action of the steering, it is worthwhile trying to adjust this for racing use. A strong castor action is very tiring on a race circuit and makes precise steering difficult.

Racing cornering

When driving around a corner at high speed centrifugal force will be acting upon the vehicle mass. If the vehicle speed around the bend is too fast, the centrifugal force will overcome the tyres' adhesion to the road surface and the car will leave the intended line of steering. The wider the arc that the vehicle describes when driving around the corner, the greater the speed that can be attained without overcoming the tyres' adhesion to the track surface. The higher the speed at which the car exits a bend, the faster the speed can be built up over the following section of the track.

There is one fastest route through every corner on a circuit. This route is called the 'racing line', and all drivers will use it, unless the track conditions preclude it because, for instance, oil has contaminated the surface on the racing line.

Every corner or bend has an apex. In racing driving terms there is a 'theoretical apex' (the geometric apex) and a 'practical apex'; the practical apex is the one a racing driver aims his car for. The illustration shows a car cornering (at X speed), using a path which just clips the practical apex. This is a smooth curve and the driver should be able to apply full throttle just before, or at, the apex; thus accelerating cleanly out of the bend. Taking the same bend at the same speed, but following a path clipping the theoretical apex, the car would leave the track on exiting the bend; unless speed were reduced and steering corrections made. All of this would greatly reduce the actual speed through the corner.

Multiple bends or corners require a different approach; the important factor being to achieve the fastest exit speed at the last bend exit in the sequence. A racing line slower than the perfect racing line is taken through the first bend, or bends, in order to achieve a perfect line through the last of the multiple bends. Sometimes the layout of a combination of bends can be approached as if it were all one bend, with a racing line sweeping clean through them all. Usually, the prime consideration is given to the last exit in a chain of bends. The racing maxim for cornering is: slow in – fast out.

Oversteer and understeer

Oversteer: When the adhesion of the rear tyres to the road surface breaks before that of the front tyres does. Or, when the rear end of the vehicle tends to steer deeper into a corner than the driver intended to steer the vehicle: the car pushes the steering tighter than intended round the corner.

Understeer: When the adhesion of the front tyres to the road surface breaks away before that of the rear tyres does. Or, when the car takes a wider line around the corner than the driver intended to steer: the driver keeps turning more steering lock on to try and get the car tighter round the corner. Every car will exhibit each of these traits

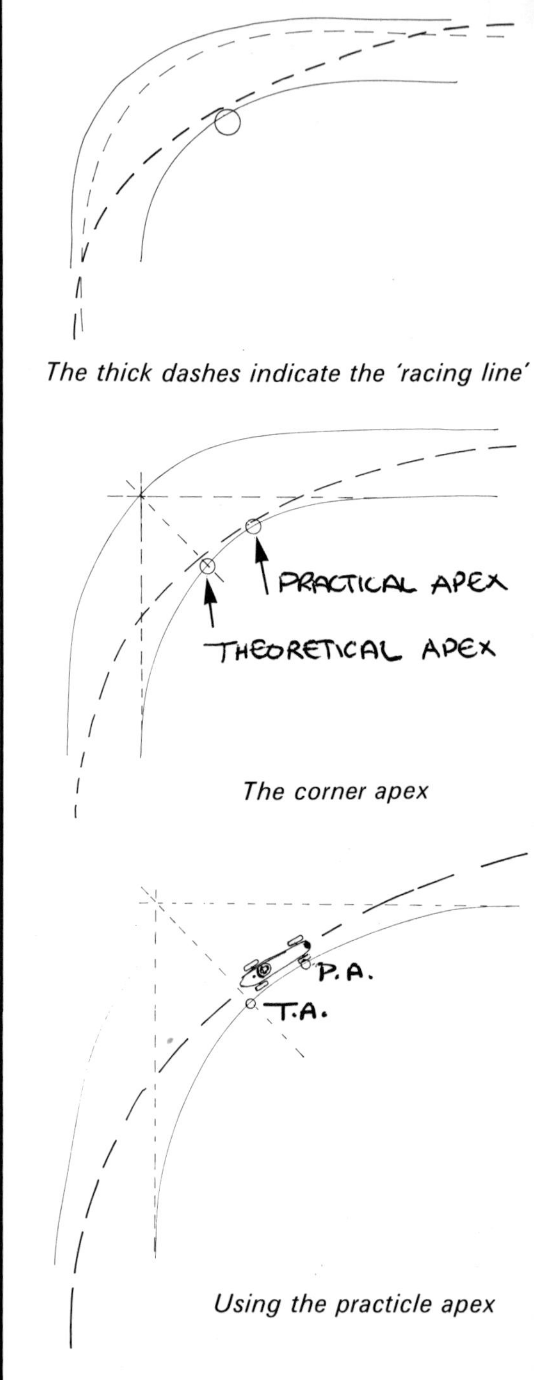

The thick dashes indicate the 'racing line'

The corner apex

Using the practicle apex

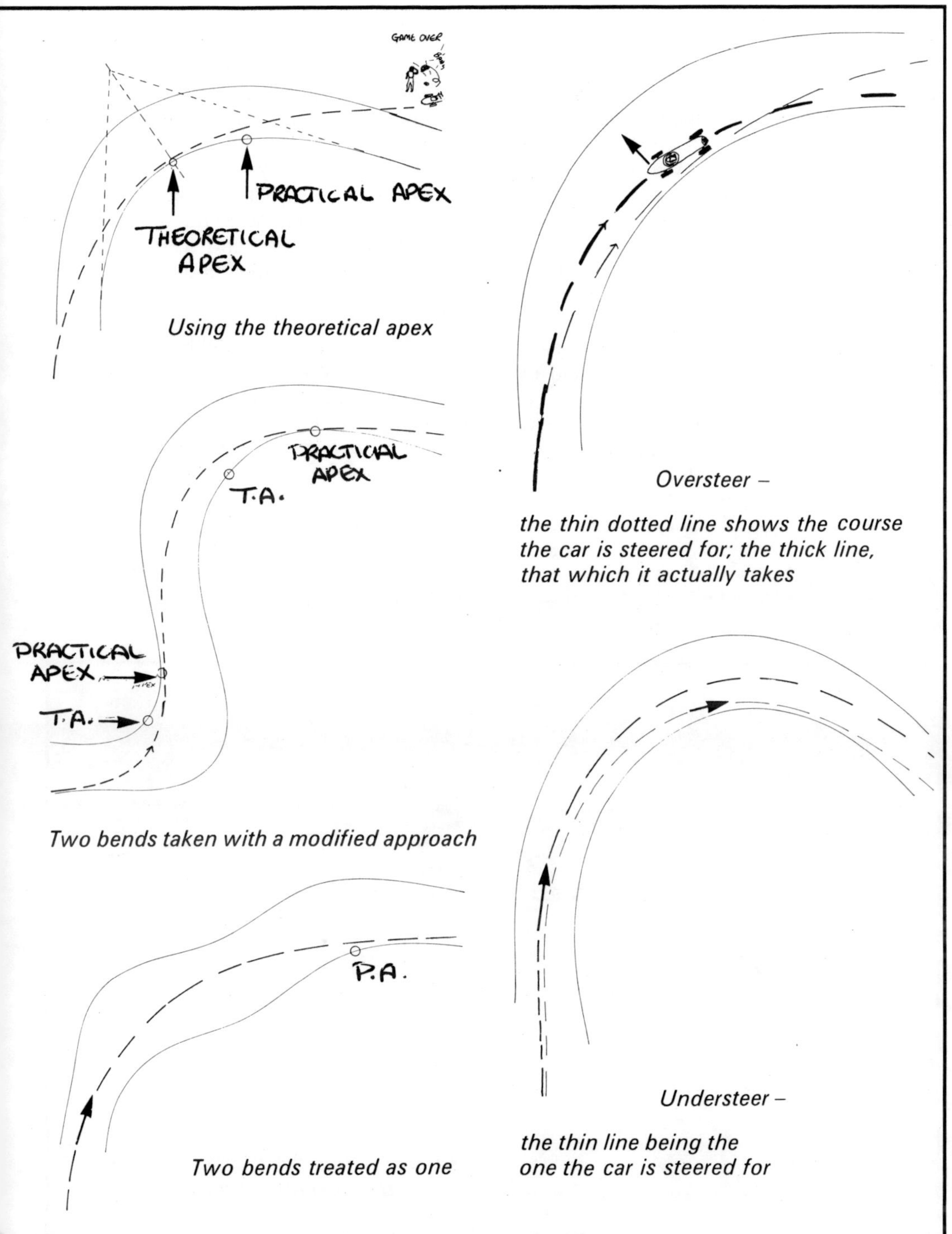

Using the theoretical apex

Two bends taken with a modified approach

Two bends treated as one

Oversteer –

the thin dotted line shows the course the car is steered for; the thick line, that which it actually takes

Understeer –

the thin line being the one the car is steered for

under particular conditions. Although some designs are predisposed to understeer, such as front wheel drive, driving technique, road conditions, and corner entry speed can all give rise to either phenomenon with a vehicle that has no design predisposition one way or the other. A racing driver should be aware of both conditions, should learn to recognise when a vehicle may enter such a state, and should drive prepared to make whatever adjustments to his technique may be necessary to neutralize them.

Oversteer, with the rear end of the car sliding a little on bends, is often used by drivers as a quick way of getting a car around a tight corner. When the rate of slide becomes excessive, opposite lock (steering opposite the direction of turn) must be applied to reduce the rate, and thus maintain control. Successfully controlling oversteer by applying opposite lock lies in judging how much steering lock to apply; and when to take it off: the necessary experience can only be gained with practice. Attempting to turn a corner, at too high an entry speed, produces oversteer.

Understeer. It is not possible to control serious understeer by making steering corrections. If a car exhibits serious understeer repeatedly on a particular bend, then a modified driving approach to that corner is indicated. Moving the practical apex forward on the bend will leave more room on the track at the bend exit, allowing for the line that the car will take.

Some front wheel drive cars, for instance the Mini, require a deliberate driving approach to work with, rather than against, the car's inbuilt handling at speed. A momentary lifting of the throttle just after the entry to a fast bend will provoke an oversteer response; the car flicks round to a tighter heading into the corner. Power is then re-applied and the car takes the required line through the bend: very quick responses are needed to make this manoeuvre work well.

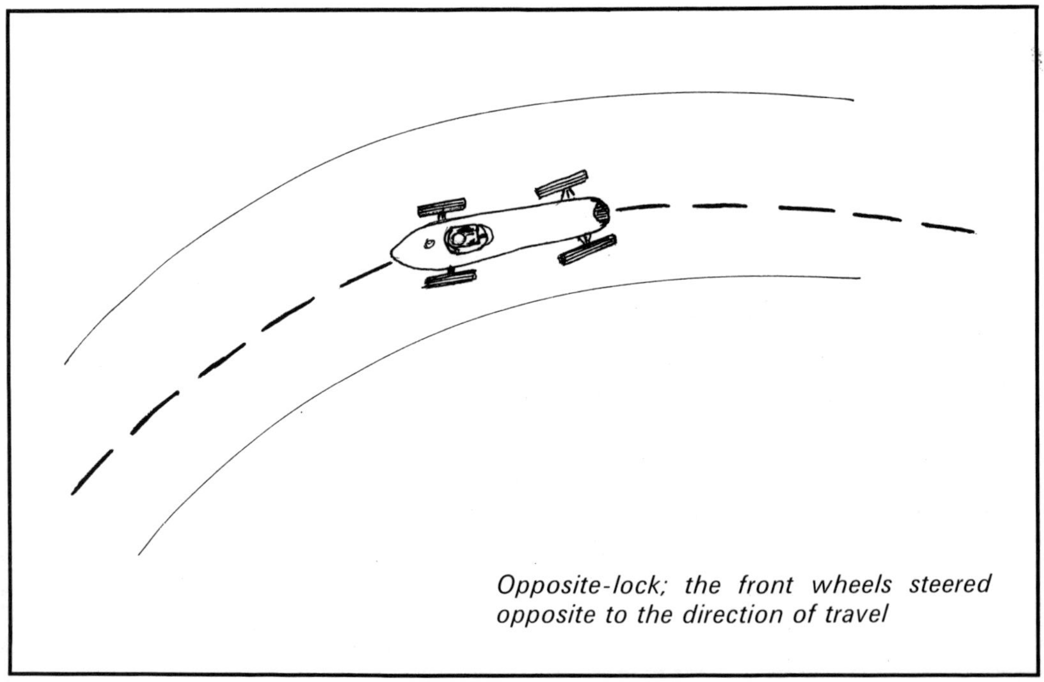

Opposite-lock; the front wheels steered opposite to the direction of travel

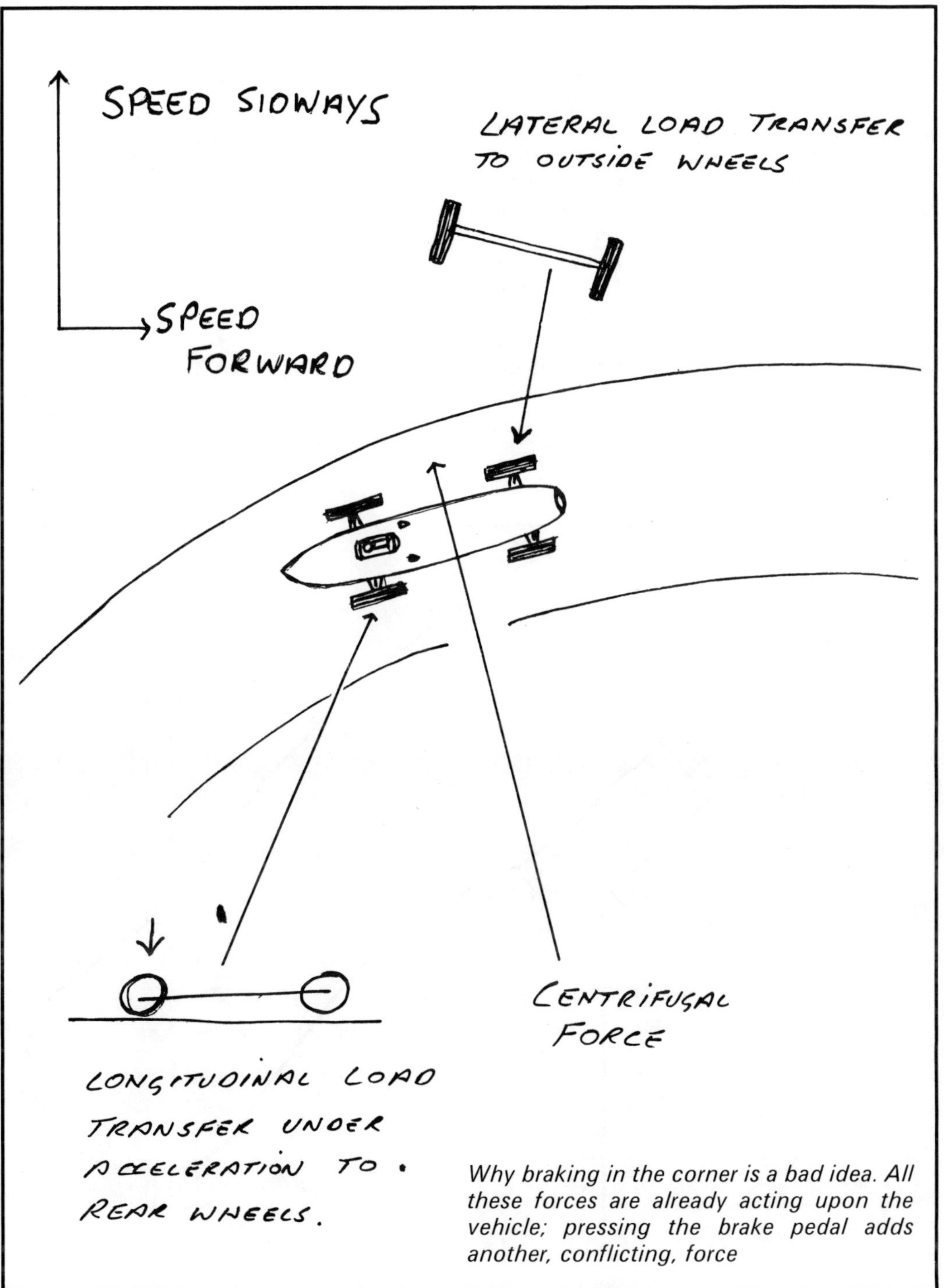

Why braking in the corner is a bad idea. All these forces are already acting upon the vehicle; pressing the brake pedal adds another, conflicting, force

Gearchanging

Fast gearchanges require the use of a 'heel and toe' technique where the right foot is being used to control simultaneously the accelerator and brake pedal whilst changing down the gears. The accelerator is blipped whilst braking and the clutch held in, this helps to match the rotational speeds of the lower gear clusters.

If the size and arrangement of the brake and accelerator pedals make the heel and toe manoeuvre difficult, the pedals should be modified. The simplest way to modify the accelerator is to fit a large flat add-on pedal as sold in speed and accessory shops.

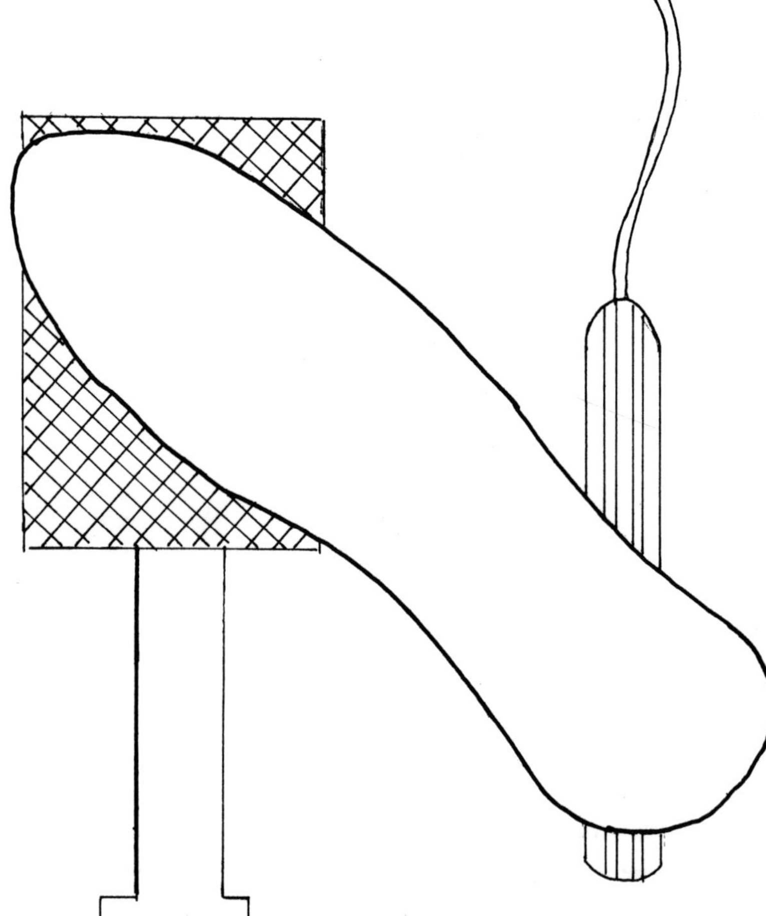

Accelerating and braking

These two opposite conditions are considered together because of their direct inter-relationship in achieving a fast lap time.

In practical terms, it takes twice as long to accelerate to a given speed, than it does to decelerate from it. As a hypothetical example: if a car takes 400 metres to accelerate from 30mph to 80mph, it will only require 200 metres to brake from 80mph to 30mph. The extent of the difference between the acceleration and braking distances is, of course, affected by the efficiency of the engine and braking system; but as a general guide rule it holds true.

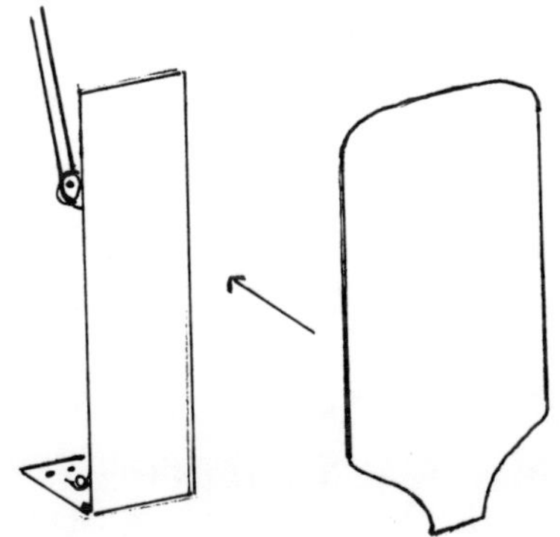

If the accelerator pedal is not big enough – bolt a big plate on it

Over indulgent use of the brakes seriously affects that important few seconds of a lap time that separates the man at the front of a race grid from the man at the back. During a race the driver must think ahead, positioning his car on the track so that the minimum use of the brakes is required.

Part of the process of becoming familiar with, and learning, each circuit is learning where the braking points are on the track. These points are the places where, just before a turn, the driver should brake and change gear. Confining the use of brakes (baring emergencies) only to these points, gives the driver the maximum length of the track available for acceleration. The point at which braking for a corner starts should not be left until really hard braking is required (late braking) because overhard use of the brakes can unbalance the car and lead to erratic progress through corners.

There are two methods of applying the brakes. The first is the normal method used by most motorists; simply to press the brake pedal. The second method is called 'cadence braking' in which the brake pedal is applied and released rapidly. In effect the turning wheel is slowed repeatedly for a fraction of a second at a time. With practice, cadence braking can be developed until in effect, it works like a manual anti-lock braking system. The technique requires sufficient speed and finesse of brake pedal application that with each application the wheels are braked as hard as possible but without actually stopping their rotation and so breaking the tyres' grip on the road surface.

Acceleration includes more than simply keeping the accelerator pedal pressed down. The racing driver must develop throttle control. Careful control of the throttle, and engine power, influences steering and makes the best use of the performance available.

Even an untuned engine can give very good performance, if the throttle is controlled to keep the vehicle in motion and operating within the best power band of the unit. With a highly tuned engine the effect of throttle control is even more pronounced. For instance, a fully race-tuned Ford pushrod engine will run up to 9000rpm, with the power band being between 6000rpm amd 8500rpm. Below 6000rpm these engines have little useful power. To race successfully, the driver must operate the car, carrying out all gearchanges and manoeuvres whilst keeping the engine in the power band. In motor racing parlance this is called 'keeping it on the boil'.

Learning to select the right gear for each particular corner on a circuit will enable a driver to keep the engine operating in the power band.

Improving your lap time

The key to fast lap times, and improving them, is consistency. The driver's first aim on learning a new circuit must be to establish a consistent route around the track, and then perfect his driving of that route so that following it is combined in an easy rhythm with the car.

When a consistent lap time has been established, it is easy to see what effect any variation of the route or driving technique will have on it. Any difference that produces a repeatable improvement can then be incorporated into a new route or rhythm on the circuit. Eventually, improving on a lap time then becomes a process of small increments of improvement, which eventually add up to a significant difference.

An amateur racing driver can only attain a certain standard of driving by practice and experience; beyond this (unless especially gifted) the knowledge of experts is required. There are a number of excellent books on the subject of the theory and practice of race driving; any driver wishing to improve his technique will find the time spent reading them well repaid.

10
Winning techniques

AMATEUR MOTOR RACING is as much concerned with the car – the pleasure of owning, maintaining and driving them – as it is with actually driving in a race to win. Personally, the simple pleasure of driving on a racing circuit is my reason for taking part; but then my idea of The Creation would be Adam and Eve in a Bugatti Type 35. Some drivers race to pit their nerve and ability against others and the challenge of the circuit; for them the car is secondary.

There are three factors which can combine in one competitor to produce a winner, or that separately, with the right combination of conditions, can lead to a driver winning a race. These aspects are: the mental attitude of the driver (coupled with ability); the competitiveness of the car (versus the other vehicles in the race); and the reliability of the car. It follows that a deficiency in one aspect can be overcome by a superiority in another, but unless the 'will to win' is present the result will rely on that undefinable quality – luck.

The mental attitude to driving in a race to win is an ability that can be acquired. This attitude is not a mental state that can be quantified and measured, neither is it definable; it is different and personal to each driver. But many statements used to describe a racing driver's attitude to driving sum it up very well: to drive in anger; controlled aggression; driving on the limit.

Such an attitude applied alone does not produce a winning driver; but applied on a base of solid experience it can.

The two essential solid bases on which to build experience that could make you a winner are, the ability to learn the best route around a circuit, and consistency of performance.

Each racing circuit has one fast route around it: one best line through each of the bends, one best place to change gear in the right combination and brake for each bend, which will extract the maximum performance from the car. Learning the circuit means learning and remembering the fastest route around a track. As a novice driver you can follow and watch the route the fastest drivers use. As you gain experience you will develop the ability to learn a circuit quickly.

All racing circuits have individual features which you can memorise and use as marker points on the track. A post or fence can be used to mark the turn-in point for a bend, or the contour of the track surface at some point can be used to remember a particular action, such as braking. You must learn and remember a circuit before you can drive around it at a consistent lap time. Consistent lap times follow learning the best route around a circuit, and making the best use of your car's performance on each track. When you have developed the

ability to produce a consistent lap time you can turn your attention to trying to improve on them. You can only improve on a consistent ability, since otherwise variables such as luck or following a line you cannot repeat through a bend, make meaningful measurement impossible. Consistency also comes with familiarity with your car, when you are able instantly to settle into a rhythm in your driving; this rhythm can be aided by your driving style.

The competitiveness of the car is an important factor which can be influenced as much by the standard of preparation as the amount of money available to spend on it. Even the most expensive components will not function properly if poorly assembled and maintained. In most amateur racing formulas equivalency rules ensure that every driver competes against fair competition for his car and accordingly the quality of preparation is paramount.

Reliability is also dependent upon good preparation and an old racing maxim states, 'before you can win you have to learn to finish'. This applies to every stage of an entry to a race and more than one driver has made an excellent start from the race grid only to run out of petrol on the second lap. A check list of items on the car to run through before the race can ensure this does not happen.

Another factor in vehicle reliability is maintenance. Particularly with racing cars it is unwise to assume that because a component works that it is not worn. Regular inspection of the entire car in between races is essential, and in time a list of which components wear fastest can be built up. Fellow racing drivers of the same type of vehicle as yours are a good source of information about sudden failure of components: they are always willing to share such knowledge.

Driving style

There are two distinct methods of driving, both of which have their adherents and detractors; who each have valid reasons on their side.

The step method. With this method of driving it is held that each operation which has an effect on vehicle handling and stability should be carried out separately.

Proponants of this method maintain that, for instance, braking and steering into a corner at the same time can combine to upset the balance of the vehicle, resulting in unpredictable handling. They also maintain that the step technique is a safe method of vehicle control, since the driver concentrates on only one operation at a time, rather than combining several operations at once.

The step method maintains that operations should be carried out in the most favourable conditions for vehicle stability; thus, all braking should be carried out and completed whilst the vehicle is travelling in a straight line. On approaching a bend the step technique would be: estimate distance to bend and gear required – change gear down, whilst in a straight line, brake down to speed required to enter bend – finish braking, turn steering into bend, maintain throttle and gradually accelerate from apex of bend, steering correctly on exit from bend, accelerate, gearchange up.

Police driving instructors use the step technique and it is an important part of police high speed driving courses. Police driving instructors with many years experience of teaching pupils of varying natural driving abilities have found the technique to be an aid to ensuring a uniform level of competency and safety in high speed driving.

The blend method. In this technique the aim is smoothly to blend all the operations of controlling the car so that its handling is held in balance. With the blend technique braking, steering and gearchanging may be carried out at the same time.

Using the blend style the approach to a bend could be: at braking point for bend, brake and change gear simultaneously, steering into bend, perhaps continuing to brake gently, then making the transition to acceleration at the apex whilst changing gear upwards before exit of bend.

The important feature in the blend technique is that the vehicle's transition from one state (straight line) to another (cornering) is accomplished in a smooth flow: clumsy sudden changes, such as stabbing at the brake whilst on a bend will destroy the vehicles's stability.

The blend technique requires conscious practice because it is difficult to achieve smooth control of the vehicle, whilst altering its state of balance in several ways at once. However, most professional racing drivers use the blend method since at high speed on many racing circuits there is not time, or room, to adopt a step approach.

These two approaches to a style of driving can be considered as separate methods, because in order to achieve a consistent rhythm in your driving you will need to adopt a consistent method; and practice and refine it.

As I am unlikely ever to win a race I have asked three top drivers for their comments on winning to conclude this chapter:

Derek Bell

To win you must have the determination within yourself to do so – but – your mental attitude must be flexible and strong enough to encompass defeats; without affecting your positive attitude toward winning. Nobody wins all the time; you can lose for a hundred reasons; but if you have the ability and determination – none of them means you will not win the next time. NEVER – forget the people who supported you in the early days on your way up; not just the big ones but the little

Even top professional drivers enjoy amateur motor racing. On the left Le Mans winner Derek Bell with Graham Bayley (beard) and other amateur members of the Alfa Romeo Team at a Donington 'Historic Race' weekend

ones too: everyone who helped, in any way, made it all possible.

Stirling Moss

'Show me a man who is second and I will show you a loser.'

James Weaver

'No matter what you have, make one hundred per cent of that. Any car that is as well prepared as it is possible for it to be will stand a good chance against a more expensive or sophisticated car, whose driver is not totally committed to his racing.

My other piece of advice is: do not get 'psyched-out;' because other people have more money or newer cars, do not think yourself into losing before you have even started'.

11 How race meetings are administered

APART FROM the officials and members of the motor racing club you elect to join, you will also have contact with the following officials who are in charge of, and organize, race meetings.

The chain of command at a race meeting

The clerk of the course

The clerk of the course is the man in charge of the whole race meeting. Although many of his duties are carried out by delegates, it is the clerk's responsibility to ensure that all the RAC rules and regulations governing a meeting are followed. These include the

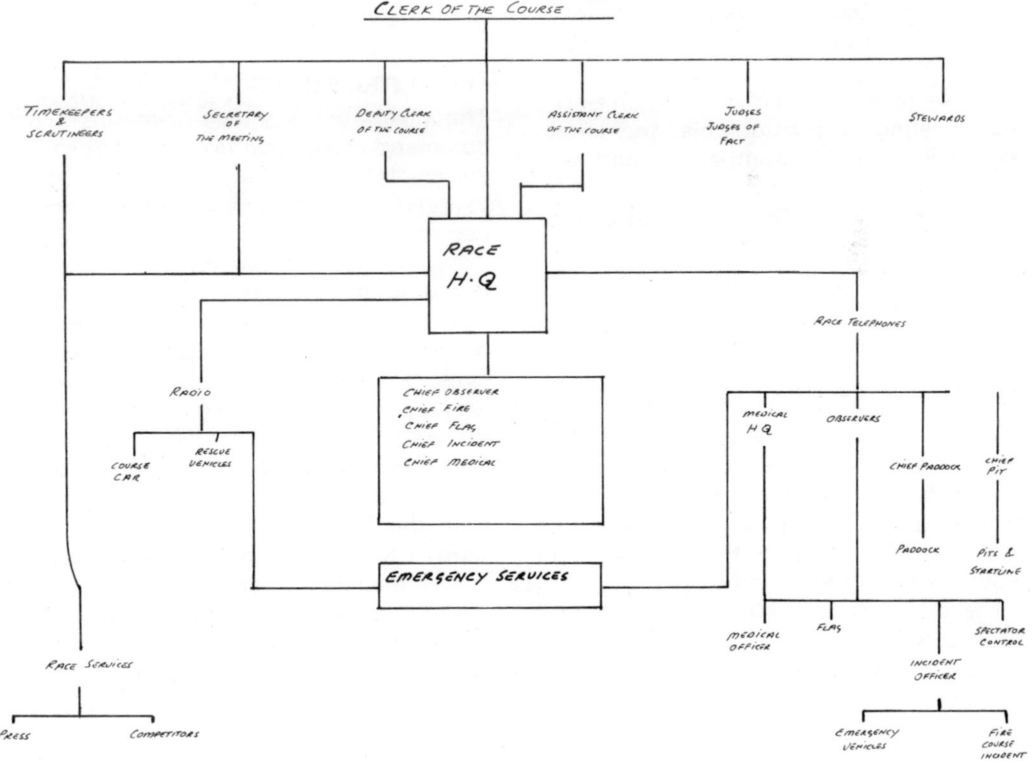

checking of competitors' racing licences and medical certificates; signing the licences after a race for upgrading records; ensuring that scrutineering, timekeeping and marshalling are properly carried out.

The clerk of the course has absolute authority over a race meeting. He may exclude from the competition any driver reported during race practice for careless or reckless driving; he may also impose such penalties as a recorded ten seconds added to a competitor's lap time for less serious infringements.

Racing drivers competing at any circuit for the first time will be instructed to report to the clerk of the course for a briefing about the circuit and any important features to be aware of.

The secretary of the meeting

The secretary is responsible for all the administrative functions of the organization of a race meeting. These include the issuing and processing of race entries, issuing organizational notices to officials and marshalls, and the issue of practice time sheets and race result sheets.

The secretary's workload in organizing a race meeting is considerable, the letters and notices sent to competitors and race officials may be running into several hundred. Accordingly, the wise competitor sends in his race application at the earliest permissible date, and waits patiently for the postal service to bring his acceptance form.

Stewards of the meeting

The stewards of the meeting are the first stage of the judicial body of motor sport. One of their functions is to submit a report of every race meeting, including all results and any comments on the conduct and organization of the meeting.

The stewards can adjudicate upon any dispute or protest at a meeting, and under certain circumstances are empowered to halt any competition. The stewards hold the general power and authority to enforce compliance with the RAC regulations and any supplementary regulations.

Judges of fact

The judges may be appointed by an organizing club, or the RAC-MSA, to adjudicate on any incident or question of eligibility.

The judges are appointed from, and act in, fields in which they are experts or in which they have considerable experience. An example is the Chief Scrutineer and members of the RAC Technical Commission, who act as final judges in questions of vehicle eligibility.

*　*　*　*

The following section describe the work of some of the other personnel needed to staff and run a race meeting. All of these jobs are carried out by volunteers and such work is a good way to become involved with motor racing other than owning and driving a racing car.

Race marshalls

The illustration of a race meeting chain of command shows that from box 1 (containing chiefs) marshalling is then divided into various areas of command for important controls; from there, other sub-divisions exist. For marshalling purposes, a racing circuit is divided into zones, so that each zone can be fully monitored and controlled by a team, reporting back through the chain of command. Most importantly, this ensures that when emergency services are required they can be directed at a second's notice to exactly where they are needed.

What is a marshall? Quoting directly from the British Motor Racing Marshalls Club (BMRMC) training notes booklet:

'A marshall is usually a motor sport enthusiast who lacks the finance and/or ability to compete but who is, nevertheless dedicated to the sport.

A marshall can come from all walks of

life – and does – to discover a common bond between all marshalls.

A marshall has a degree of insanity in his make up. This is essential, to enable him to stand about from dawn to dusk, probably in atrocious conditions, in pursuance of his duty.

A marshall therefore is no ordinary person!'

The following details will give a clear idea of exactly what some marshalls' jobs are.

The Marshall – ready to leap into action at a split second (courtesy Barry Foley)

Course marshalls

This is one of the first assignments on beginning marshalling to allow the new recruit to gain experience. The duties include inspection of the circuit, clearing loose stones and debris, reporting track conditions to the Observer, assisting at incidents.

Fire and incident marshalls

This is a sub-division of course marshalling. The marshall is allocated to a position on the circuit with a fire extinguisher. The fire marshall deals with accidents on his section of the track. The Incident Marshall, deals with all other aspects of an incident.

Flag marshalls

The flag marshall is the drivers' vital aid to safety during racing. The marshall carefully observes all the cars during racing and indicates problems to drivers on the circuit by using different coloured flags. For example, a blue flag is waved at a driver when a following car is coming up to overtake him.

In the heat of competition it is easy for a driver to concentrate only on the road ahead, becoming oblivious to the rear view in his mirrors: right at the track side the marshall's warning flag is unmistakable.

By the very nature of the job the flag marshall must be quite knowledgeable about motor racing, and race driving in particular: the flags must be used to inform drivers of real perils, not imagined ones.

Observers

The observer is the eyes and ears of the clerk of the course; the most senior marshall in each sector or zone. The observer directs operations if an incident occurs and reports infringements of the racing regulations to the clerk of the course.

The on-site chain of command at a marshall's post is shown in the illustration.

There are many other marshalling duties, including: Pit Marshall, Startline Marshall, Time Control and Race Telephones Marshall.

Motor racing marshalls' clubs

Readers wishing to become a marshall should apply to:

The British Motor Racing Marshalls Club. Mr Mike Oxlade, 11 Deverill Road, Stoke Mandeville, Bucks.

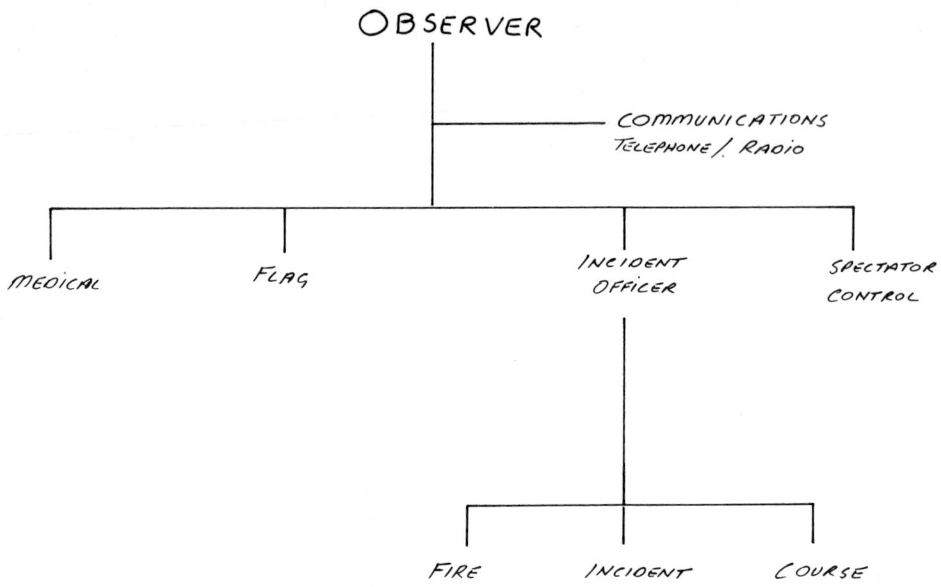

OBSERVER

- COMMUNICATIONS
  TELEPHONE / RADIO

- MEDICAL
- FLAG
- INCIDENT OFFICER
  - FIRE
  - INCIDENT
  - COURSE
- SPECTATOR CONTROL

The Observer's Post

Watch for the flag marshall's signals

The BRSCC and BARC also have sections of their clubs especially for motor racing marshalls.

British motor racing marshalls are highly thought of throughout the world for their knowledge of the sport, their professionalism and their dedication. British marshalls are often requested to help marshall important racing championships abroad.

Scrutineers

The scrutineer's job is to examine each competitor's car before practice is allowed to begin and certify that it is safe and in a fit and proper state to race; and to examine competitors' cars to ensure that a vehicle is what it purports to be. That means that when a car is entered as a 1000cc engined Classic Saloon the scrutineer will establish that it does, in fact, have a 1000cc engine and not one of 1500cc.

The scrutineer is also the official who decides questions concerning whether a vehicle conforms to the construction and technical regulations.

The scrutineers's job requires a good knowledge of mechanical engineering: to be appointed to the senior grades, formal engineering qualifications are required.

Scrutineers are graded as groups 1, 2 and 3, with a limited number of chief scrutineers. Officials of this nature receive a small fee from the race organizing club using their services, plus a travel allowance. Readers interested in scrutineering should contact an RAC listed scrutineer who deals with the type of vehicles or club of their choice.

Timekeepers

Timekeepers for race and speed events are licensed by the RAC-MSA. They are listed in two categories; those using manual timing equipment, and those using automatic timing devices.

The timekeeper's work requires considerable application and concentration, coupled with absolute accuracy. Timekeepers progress in their duties by continual practice, and start off as assistants to graded officials. To become a timekeeper an application must be made to the RAC-MSA for appointment, and be accompanied by a recommendation from a club, or a listed timekeeper for whom the applicant has carried out assistant timekeeper duties.

12

Sponsorship

The amateur's position

FORMULA ONE motor racing is now a business within itself, and has grown to its present stature through the shrewd hard work of those who control it. With television rights, huge crowds, and the impact for a product with its name featured on a Grand Prix car, it has become self-financing: a separate world within motor sport. With the cost involved in Grand Prix racing today that professionalism is essential for its survival.

In the lower echelons of motor sport, during the best years some sponsoring companies were awash with money – money that would otherwise go in tax. So, through the sixties and seventies companies could dabble in racing sponsorship, writing the cost down to advertising and offsetting part in taxation savings. In the last few years this has changed, as recession affects profits, companies need to justify every item of expenditure, and see a return on it. This has filtered through to amateur motor racing, too, where once there were a few genuine benefactors willing to help a club or aspiring driver.

So, where does that leave the amateur racing driver today? Back where he was in the nineteen-fifties – doing it all himself with a little help from his friends.

To gain any financial help with your motor racing you must make the right kind of effort to impress such friends.

To persuade anyone to sponsor you with money or work you must offer something in exchange. That something is association with your 'image'; the sponsor must feel that he will gain something in return, tangible or not, from his association.

The first step is to create this 'image' – as something that is attractive, and that people will want to be associated with. Simply asking for financial help with your racing offers no benefit to a sponsor – in effect it is asking someone else to pay for your pleasure. But a small business associating with an image of excitement and vitality could feel that their business will be identified with that image by potential customers: what you are offering is of mutual benefit.

Creating an image

Very few amateur racing drivers build a racing car and take part in the sport entirely alone. Usually a friend or friends are involved, who take on the duties of mechanic, whilst wives or girlfriends often become the team manager and organizer. The group involved with the car form the 'racing team': the first stage of creating an image.

An image of this nature needs a name, a persona, to focus upon; so careful thought should be given to finding a suitable name

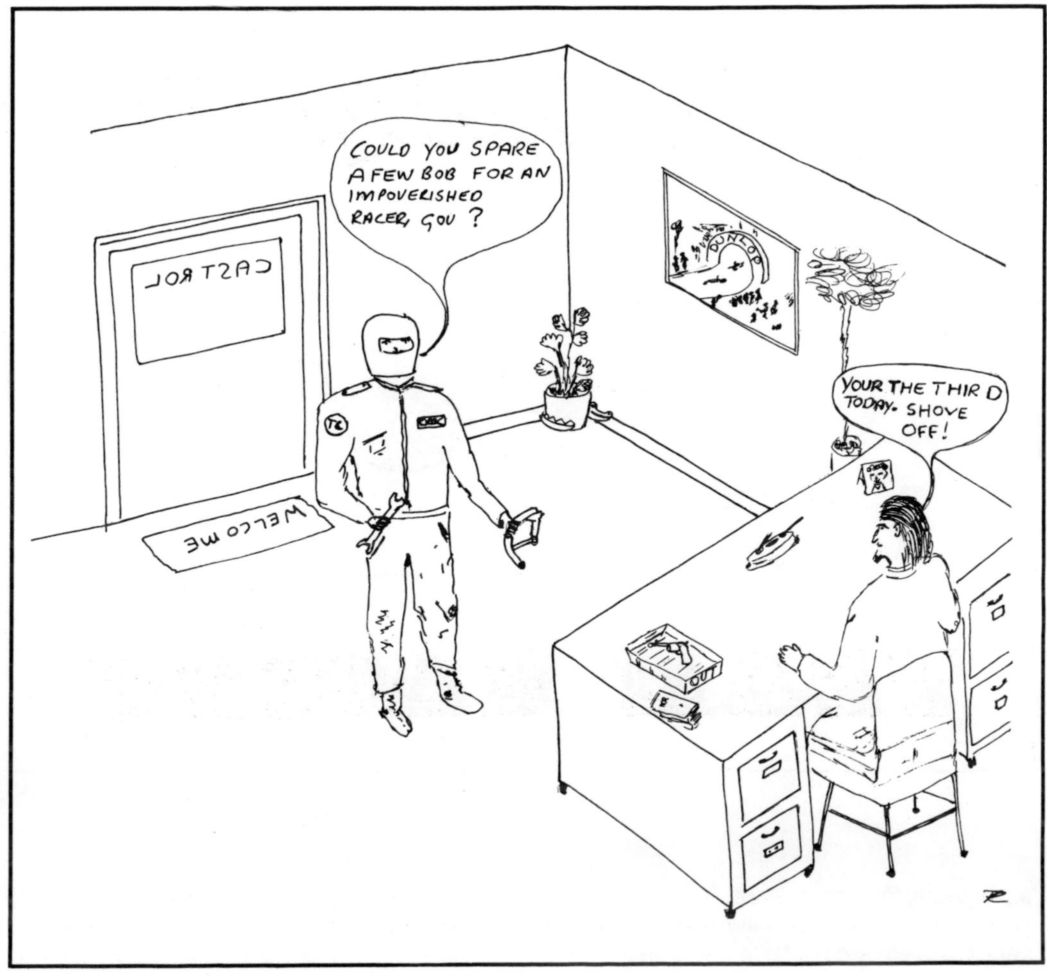

which will help to create an identity for the team. Many drivers simply use their own name for the team; but a more interesting name (for instance – Jester Racing Team) may help give the racing team an identity of its own.

A logo or symbol is also useful to give the team an identity: simple cloth badges and inexpensve stickers can be obtained. Matching overalls, or sweatshirts, with the team badge on, make a good impression when worn for photographs of the team with the car. Stickers with the team name and logo used on the car also help to present the team vehicle as an image.

The racing car should always be as clean and presentable as possible. Potential sponsors will usually want to see the vehicle, and a scruffy car creates a bad impression.

Photographs of the team and car, preferably with a racing venue in the background, are essential for showing to potential sponsors. When you approach a businessman and show him a photograph of the team you present something tangible to identify with.

Even more effective than a photograph is

The author's team logo

a printed information sheet incorporating a picture of the car and team: the information could be a calendar of events to be entered, or past successes together with information about the team. The addresses of companies producing this type of printed material inexpensively can be found in *Exchange and Mart.*

A further step in creating a saleable image for the team is to obtain some mention of your activities in your local newspapers. Local newspapers are always on the lookout for such items of interest, and a personal visit to their offices should be made. If the newspaper staff do not have the time to call and see the vehicle they will usually accept your own photograph and notes as a basis for a small article. If you have a humorous story about the car or its construction that can often aid publication and increase the amount of coverage the newspaper will give to your team and the car.

Local radio stations are another source of publicity for your team, but before approaching them you should collect some notes on interesting aspects and stories about the car and team: as no pictures are involved a simple description of the car is hardly interesting enough for broadcast.

What you have to offer a sponsor

When you approach a potential sponsor you should have a clear idea of what you can offer him in return: consider how it may aid his business.

The first basic benefit you can offer a sponsor is to carry stickers advertising his company on the racing car. In amateur motor racing this form of advertising is of very limited actual benefit, since no television coverage or major exposure of the advert is possible. However, pictures of the car in local newspapers, with the stickers prominently displayed, can give some good local exposure, which could be useful to companies which derive much of their business from the immediate locality.

Depending on the type of business, the racing car itself may be a useful attraction for promotional purposes. Businesses such as a record store or a garage may be able to generate interest in visiting their premises by having the racing car on display, such as a feature spot inside a showroom or a one day exhibition outside the establishment.

The car and team may also serve as an attraction in conjunction with a company display at a local exhibition or gala. For such venues the driver or whole team should be prepared to be on hand to answer questions about the car, and assist in promoting a sponsor's products or services.

Promotional material featuring the car and team with a sponsor's staff or products is another possible benefit. For these purposes much larger stickers or company advertising than normally used when racing can be applied to the car: this can increase the promotional value of the vehicle since the size of advertising stickers permitted under RAC racing regulations is strictly limited.

Although amateur motor racing is rarely featured on television, video taping of races by club photographers and individuals is now common. Video cameras can be hired quite cheaply and a video showing a sponsored car in action is quite

Drawings such as this, used as a mail shot, can interest companies before you call

simple to take. There are a number of companies who will edit and compile a promotional video film from a collection of amateur video tapes. This could be another useful benefit to offer a sponsor.

Included with a race entry are five circuit entry tickets and two passes for cars; additional entry tickets can be purchased by an entrant at a discount rate. You can offer tickets to sponsors or potential sponsors so that they can see the car in action; this is a good way of making them feel involved with the team and car.

If you invite sponsors to attend a race meeting it is very important to entertain them there. Make the effort to meet the sponsor at the circuit, show him around the paddock and the cars, explain what is happening, show him where the cafes and stands are, and speak to him before and after practice and the race: the effort will pay dividends later.

Lastly, sponsors can have their name printed in a race programme as the 'entrant' of a race car if they purchase an entrant's licence from the RAC. As well as the entrant's name, which can be a company name, a few other descriptive words are allowed in the programme and this is another form of advertising.

What you can gain from a sponsor

Straightforward financial help with racing costs is the most difficult form of sponsorship to find: careful attention to and presentation of a saleable image and perserverance in calling on possible sponsors is the only way to find one. However, help in the form of assistance with work or parts for the car can be just as valuable, and is easier to obtain. Welding or machining work, fuel, tyres, are all expensive to purchase as an ordinary customer; but supplied at trade prices are a great help to the amateur racer, and a negligible loss to the sponsor. If over a period of time the sponsor comes to feel that his company is gaining some benefit from the association with racing, the arrangement of supplying parts at cost price can progress to their supply at no charge.

The author's sponsor, Jerry Randell, outside his hairdressing salon 'Root 66'. Photos like this in the local press can help the sponsor and team

The Classic Car Team of owner/ constructor Andy Anderson, driver Nick Amey and team manager Wendy Markey

Calling on potential sponsors

Any local company may be a potential sponsor. Garages and companies connected with the motor trade are usually the first choice, although they may well have been canvassed before. Companies which actually carry out most of their business locally could usually benefit from local advertising ; this aspect of what you have to offer should be emphasized.

British Telecom Yellow Pages and local directories are a source from which to compile a list of companies to approach.

A folder of photographs of the car, press cuttings about the team, and similar material, is a help to show the potential sponsor what it is you are asking him to support. You should also be prepared to take the racing car to his premises to show him: before calling on sponsors make sure the car is in a fit state to be shown to them.

Calling unannounced on business people from whom you hope to seek sponsorship can be counterproductive; they may well be busy and thus, unreceptive. A good approach is to telephone first, making it clear that you will only take up a few moments of their time: when asking for an appointment offer several alternative times and dates when you could call, this makes it more likely that someone will agree to see you. It is more difficult to get someone to offer a time for an appointment that they are not keen on making.

Make your first requests to a sponsor for cash, work or parts, small. Asking for a whole list of items will make the sponsor tot up the cost mentally; he may be put off.

Particularly at first meetings, offer tickets to the next race: this shows that you are willing to offer some benefit to the sponsor straightaway.

RAC racing car advertising regulations

Amateur racing drivers are allowed a maximum of five stickers per car, not exceeding 55sq in each, on each side of the car. They may not be related to form a name or message: each sticker must be self-contained.

Additionally, RAC advertising permits can be purchased each year; these allow almost unlimited advertising on the car. Permits are available for race categories: restricted, national, and international. A restricted race class advertising permit costs £40.00.

Note that some motor racing clubs impose additional restrictions on the use of advertising.

Appendix 1: racing circuits and venues

This is a list of the foremost racing circuits. I have added my own comments on those at which I have raced.

Aintree

Situation. 5 miles north east of Liverpool. Tel (Clubhouse – Tuesday evenings only) 051-523-3474.

Brands Hatch

Situation. Near Fawkham, Dartford, Kent. Tel 0474-872331. This is the circuit management telephone, operating during normal office hours. Brands Hatch is, of course, world famous, and to drive on this historic circuit is a wonderful experience. For a novice it is a difficult track on which to attain a good lap time, but it has excellent facilities and is a very 'friendly' circuit used by simple amateur clubs and the world's best alike. Just the thrill of racing on such a track is enough; finishing last is no deterence.

Cadwell Park

Situation. 8 miles north east of Horcastle, Lincs. Tel 0507-84-248. Miles away from anywhere, this track is also a legend. It is a difficult circuit, with features such as its 'mountain', a sharp incline after a virtual chicane, here a car can take-off for a second, as it is powered over the top. All the amateur clubs use this circuit.

Castle Combe

Situation. 5 miles north west of Chippenham, Wilts. Tel 0249-782417. Exclusively a clubman's circuit, and thus very much an amateur racing home. The facilities are spartan, but this is more than made up for by the friendliness of everyone there. The circuit is a very good one with only one really difficult section, Quarry Bend, to master.

Donington Park

Situation. Castle Donington, Nr Derby. Tel 0332-810048. A fantastic circuit, with excellent facilities. It is a very good track to drive on, with not too many difficult sections, and plenty of grass run-off areas for real mistakes. The atmosphere at circuits is mostly made by the competitors and the clubs there on the day, but the management and officials also have their effect and here they are very good.

Ingliston

Situation. 7 miles west of Edinburgh on the A8 Glasgow road.

Knockhill

Situation. Dunfermline, Fife.

Lydden

Situation. 7 miles south west of Canterbury, Kent. Tel 0795-72926. This is a good club circuit.

Mallory Park

Situation. Mallory Park, Kirkby Mallory, Leicestershire. Tel 0455-42931. A superb little circuit which nearly disappeared when its owners put the site up for sale. The track was rescued by, and is now run by, the BRSCC and their hard-working Mrs Edwina Overend. who deals with the administration. A very friendly circuit and a good one to drive, with mastery of the hairpin bend Shaws Corner and the Lake Esses being the key to a fast lap.

Oulton Park

Situation. Near Tarporley, Cheshire. I have not personally driven this circuit, but friends tell me it is very good.

Silverstone

Situation. Near to Silverstone village, off the A43 between Towcester and Brackley, Northants. Tel 0327-857273. Another world famous Grand Prix circuit, and much used by all the amateur clubs. The facilities are good and there is always a friendly atmosphere. The circuit has its own tricks if a really fast lap is to be achieved but it is by no means difficult to drive, even for the first time novice.

Snetterton

Situation. Off the A11 between Thetford and Norwich, Norfolk. Tel 095-387-303. In a rather remote spot, this circuit is well used by many clubs, and as it is so far away from population centres, 24 hour races can be run there. I found it a difficult track on the one time I drove there, but my total lack of success there was more than made up for by the pleasant, friendly and helpful advice of the officials and organisers.

Thruxton

Situation. Off the A303, 5 miles west of Andover, Hants. Tel 026-477-2607/2696. Thruxton is a full scale international circuit and many important races are held there. It is also an excellent club venue and Ian Taylor's racing drivers' school is based there. Like all circuits it has its special difficulties, but it does have good areas of grass run-offs and is generally not too intimidating for the total novice.

Hillclimb Venues

This is not a complete list as venues are all listed in the RAC Blue Book, but the following will give an idea of those in regular use.

Baitings Dam. (West Yorks) 430 yards. 110/012191. 5 miles SW of Sowerby Bridge, West Yorks.
Barbon Manor. (Cumbria) 890 yards. 97/631824. 2 miles NE of Kirkby Lonsdale, Cumbria.
Bodiam New Farm. (E-Sussex) 700 yards. 188/783234. Bodiam Village, 10 miles north of Hastings.
Castle Howard. (N-Yorks) 440 yards. 100/710699. 5 miles west of Maldon, N. Yorks.
Cricket St Thomas. (Somerset) 193/384096. Off the A30 Crewkerne – Chard road.
Gurston Down. (Wilts) 1160 yards. 184/026252. 1 mile west of Broadchalke off A354 from Salisbury.
Great Farthingloe Farm. (Kent) 1255 yards. 179/290400. 1 mile SW of Dover.
Harewood. (West Yorks) 1090 yards. 104/338450. 7 miles NE of Leeds.
Loton Park (Shropshire) 1475 yards. 126/358143. 8 miles west of Shrewsbury.
Oddicombe (Devon) 750 yards. 202/924657. Adjacent to B3199 north west of Torquay.
Prescott. (Glos) 1127 yards. 163/985297. 5 miles NE of Cheltenham, Glos.

Shelsley Walsh. (Hereford/Worcs) 1000 yards.
149-138/721631. 10 miles west of Worcester.
Tregrehan. (Cornwall)
200/204/053532. 2 miles east of St Austell.
Valence School. (Kent) 712 yards.
188/45443. 1 mile east of Westerham, Kent.
Werrington Park. (Cornwall) 1250 yards.
201/323862. 1 mile NW of Launceston.

Wiscombe Park. (Devon) 1000 yards.
192/182941. 6 miles south of Honiton.

Sprint venues

There is also a long list of sprint venues, but many of these gatherings, like the Brighton Speed Trials, are held only once a year. If you are interested in these, your own club will keep you informed. Sprint meetings, as opposed to races, are also often held at some race circuits.

Appendix 2:
some useful addresses

The companies below are a few with which I have completed business; literally hundreds more can be found in the pages of *Motoring News* and *Autosport.*

Burton Power Products

623/63 Eastern Avenue, Ilford, Essex. 1G2 6PN. Tel 01-554-2281 or 8507. Burton supply all new Ford tuning parts, from simple uprated valve springs to full-bore race equipment. They also manufacture new camshafts and have machining and balancing facilities.

Osseli Engine Service

Ferry Hinksey Road, Oxford, OX2 0BY. Tel 0865-248100. Osseli make and supply all new tuning equipment for, Ford, BMC, and Leyland engines. They also supply complete engines and have machine shop and balancing facilities.

Autosprint

214, Livery Street, Birmingham, 3. Tel 021-236-5133. Autosprint are balancing specialists; they offer a balance 'whilst you wait' service if you can take your parts there in person.

Gemini Engineering

Sun Lane Service Station, Gravesend, Kent. Tel 0474-534779. Gemini offer engine assembly and balancing.

Kent Performance Cams Ltd

Unit 1-3, Military Road, Shorncliffe Industrial Estate, Folkestone, Kent. Tel 0303-38666. Kent make performance camshafts for practically any engine.

Jonspeed

Galley Common Service Station, Hickman Road, Nuneaton, Warks. Tel 0203-394421 for spares. 0203-3941126 for tyres. Very useful: deals in all sorts of secondhand racing engines, rolling chassis and parts, also road and racing tyres – new and used.

Nick Stagg Motorsport

Tel Chipping Sodbury 323130. Sells secondhand Mini racing parts.

Mini Sprint

Tel Leicester (0533) 702996. Deals in used racing parts.

Transpares

155 Heysham Road, Morecambe, Lancs. Tel 0524-419623. Sells tuning parts for Sunbeam and Avenger cars.

City Speed (Gloucester) Ltd

Mercia Road, Gloucester. Tel 0452-415848. Deals mostly in Ford engine parts, but has all sorts of small bits and pieces required for racing. Deals by post.

Rally Spares

Low Mill, Town Lane, Whittle-le-Woods, Nr Chorley, Lancs. Tel 02572-69245. Sells used competition parts for Fords.

Race Parts (UK) Ltd

Station Road Industrial Estate, Wallingford, Oxon. Tel 0491-37142 or 37740. Sells all sorts of small bits and fittings.

EB Spares

2 Washington Road, West Wilts Trading Estate. Westbury, Wiltshire. Tel. 0373-823856. Sells discount-priced new Alfa Romeo spares.

C.P Motors

2 Hermit Place, Rear 246 Belsize Road, London, NW6. Tel 01-625-6938. CP Motors is run by noted Alfa racing driver Pietro Caccaviello; he deals in all Alfa Romeo parts and cars.

Mitcheldever Tyres

Sutton Scotney, Near Winchester. Tel 0962-760000. Offers very keen prices on all types of tyres.

Appendix 3: special offers

Before giving details of these I must say a few words about the people and companies that have generously agreed to make the offers. Companies that make specialised racing equipment operate in a difficult situation and the volume of sales is often quite small; yet the tooling and manufacturing equipment is expensive. In the genuine racing side of their operations the vagaries of fortune that affect teams and formulas, can alter drastically the supplying companies' financial position over the period of a few weeks. On top of that, being the manufacturer of racing equipment is a very competitive enterprise, like that of the sport it supplies: profit margins are not large, the overheads are, and that makes these offers even more generous than they appear on paper. Finally, these are not offers of gimmicks to increase sales, but genuine help with the expense of getting your first racing car together and out on the track. Where the offers are to be routed through me it is at the companies' request, in order to cut down on their administration costs and make the process simpler. Please note that the publishers of this book are in no way connected with these offers and correspondence on the subject should NOT be sent to them.

(coupons appear on page 111)

✳✳✳ **Offer Number 1** ✳✳✳

**Spax Shock-Absorbers.
20 per cent discount**

Spax are offering 20 per cent discount off their UK retail prices. This offer is kindly made by Spax Managing Director Jeremy Rossiter, a racing driver himself, who is keen to encourage new drivers and with this offer is doing what he can to help.

To take up this offer you must write to me (my address is given in the queries section) telling me the make, year, and model of car for which the shock-absorbers are wanted. Enclose a stamped addressed envelope, plus two more second class stamps (these are for me to contact the Spax works and obtain a quote). You must also attach the coupon cut from this book. I will send you the details and the quoted special price and delivery dates, with instructions for completing the order.

Jaybrand Racewear.
15 per cent discount

This offer had kindly been made by Jaybrand Director Roger Hawkins.

Jaybrand are a long-established company in the field of specialist suits and clothing for racing drivers. It is their sole business, not a sideline to general overall manufacturing.

I have one of their two-layer Nomex suits together with Nomex underwear. The important thing about such kit is that not only must the material be right, but that it is also sewn, as Jaybrands suits are, with fireproof Nomex thread as well. It is useless in a fire, if all the sewed seams burn and the whole suit falls to bits.

To take up this offer, write direct to Jaybrand at the address given, clipping the coupon from this book to your letter.

Fireater Extinguishers.
15 per cent discount

Fireaters Richard Hackett and General Manager S.C. Hoskins are the two gentlemen from this company who have kindly agreed to help new racers who want top quality equipment of this sort.

Fireater have manual, mechanical, and electrical systems on offer, all using Halon 1301 gas (BTM) extinguishers. Their equipment is used by most of the British military and police forces as well as many top race teams.

To take up this offer you must write to me with details of your requirements enclosing the cut out coupon from this book, together with a SAE and two second class stamps. I will then obtain a quote from the company and send you the details.

Ian Taylor Racing Drivers School

Ian Taylor's Racing Drivers' School, Thruxton. £10 discount.

For new competitors, who feel they need some instruction before joining the fray, Ian Taylor had kindly offered a £10.00 discount off the price of one of his one-day racing driver's schools.

I must mention that the number of places on these one-day schools is limited, as are the number of days each year on which they are held. A racing driver's licence is not required. The one-day schools are also quite expensive to set up and organize and, of course, Ian needs a number of experi-enced racing drivers to help instruct the pupils. Accordingly, this offer is strictly limited to those readers actually *building a car* and preparing to enter a first race.

To take up this offer, write to me enclosing the cut out coupon, a SAE and two stamps; give me details of your car and tell me when you hope to enter your first race. I will reply and give you details of the school date available. Bear in mind that you will have to get to Thruxton circuit before 9 am.

The Queries Service

If you have any problems or difficulties with regard to building the car, or any other questions appertaining to motor racing, then write to me at the address below and I will do my best to answer them.

I cannot personally guarantee to know all the answers, but I probably know someone who does, and I will find an answer from them.

Please send me the cut out coupon with your first enquiry, and I will then put your name and address on my file, after that you can write as many times as you wish and I'll do my best to find an answer, but, please – you MUST enclose a stamped addressed envelope for your reply.

My address is:

Terry Cole,
Helms Deep,
5 Station Road,
Newnham-on-Severn,
Gloucestershire,   GL14 1DA.

Queries Service

In the accompanying letter you will find details of my query. Help!
Send to: Terry Cole, Helms Deep, 5 Station Road, Newnham-on-Severn, Glos. GL14 1DA.

Special offer coupons

To apply for any of these special offers you must:

1) Fill in the relevant coupon
2) CUT-OUT the coupon

3) Send the coupon AND a stamped, self-addressed envelope to the address on the coupon.

Spax Shock-Absorber Offer. 20% discount

Please obtain a quotation for me and send me details. My car is:

Make ....
Model ....
Year ....
Type of racing ....

Send to: Terry Cole, Helms Deep, 5 Station Road, Newnham-on-Severn, Glos. GL14 1DA.

Fireaters Extinguisher Offer. 15% discount

Please obtain a quotation for me and send me details.

My requirements are: ....
....
....

Send to Terry Cole, Helm Deep, 5 Station Rd, Newnham-on-Severn, Glos. GL14 1DA.

Jaybrand Racewear Offer. 15% discount

I have purchased the book 'The £1000 Racing Car'. Please send me details and prices of your racing driver's overalls and suits.

Send to: Jaybrand Racewear, Highbury Street, Peterborough, PE1 3BH.

Ian Taylor's Racing Drivers' School Offer. £10 discount

I am about to complete the building of my racing car and will shortly enter for my first race.

Please send me details of the school and courses.

Send to: Terry Cole, Helms Deep, 5 Station Rd, Newnham-on-Severn, Glos. GL14 1DA.